Pensions
The Hidden Costs of Public Safety

PENSIONS

The Hidden Costs of Public Safety

ROBERT M. FOGELSON

Columbia University Press
NEW YORK 1984

Library of Congress Cataloging in Publication Data

Fogelson, Robert M.
Pensions, the hidden costs of public safety.

Includes bibliographical references and index.
1. Police—Salaries, pensions, etc.—United States.
2. Fire fighters—Salaries, pensions, etc.—United States. 3. Municipal officials and employees—Pensions—United States. I. Title.
HV8143.F53 1984 352.2 84-1919
ISBN 0-231-05958-2 (alk. paper)

Columbia University Press
New York Guildford, Surrey

Printed in the United States of America

Clothbound editions of Columbia University Press Books are Smyth-sewn and printed on permanent and durable acid-free paper

Book design by Ken Venezio

To David, Lisa, Susie, Michael,
Jeffrey, Janet, and Jennifer

Contents

Preface

Several years ago, I received a telephone call from David Rothman, an old friend and professor of history at Columbia. He and Stan Wheeler, who taught law at Yale, had prevailed upon the Russell Sage Foundation to sponsor a conference on social history and social policy. Knowing that I had just finished a book about the transformation of the big-city police from 1890 to the present, David asked if I would be willing to write a paper about policing for the conference. The foundation would not be stingy, he assured me. I replied that I would be happy to write a paper for the conference, but not about policing. I had said what I wanted to say about policing in my book; and after six years of research and writing, I was tired of the subject. If he and Stan were agreeable, I would write a paper instead on the firemen's and policemen's pension problem, a problem that I had come across while doing research on the police. The heart of the problem, I told David, was that many firemen's and policemen's pension systems had gotten into serious fiscal trouble and were now putting a severe strain on the financial viability of the nation's big cities. No one seemed to know why. To find out I proposed to study one such system, the Los Angeles Fire and Police Pension System (LAFPPS). By delving into its history, I hoped to discover the sources of the pension problem. David said that the proposal sounded fine and, after checking with Russell Sage, told me to press ahead.

I chose the LAFPPS for several reasons, not the least of which was that I knew a good deal about Los Angles and could count on a high level of cooperation there. The system was in serious trouble. In 1977 it had on the rolls more than 7,700 retired firemen, policemen, and survivors, roughly three beneficiaries for every four officers, to whom it paid almost $80 million in retirement benefits. To meet current costs and to amortize prior obligations, the city appropriated $113 million, which was more than one-third of its property taxes and nearly one-half of the fire and police departments' payrolls. Despite these vast contributions, the LAFPPS had run up a deficit of $1.9 billion, close to nine times the payroll of the two departments and more than one-sixth the assessed value of the city. Nor could the system's fiscal plight be explained by mismanagement or skulduggery. The staff was extremely capable and the actuaries highly qualified. The records were well kept, the reports informative, and the funds invested in a prudent manner. The system had been free of scandal since the early 1940s. It also shunned the questionable practices that have discredited other public employee pension systems in recent years. It did not encourage officers to retire in order to escape disciplinary proceedings; nor did it rubber-stamp applications for disability pensions or compute retirement allowances in such a way that some members earned more on a pension than on the job.

Underlying my analysis of the LAFPPS were several distinct though closely related questions, the answers to which would enhance our understanding of the pension problem not only in Los Angeles but in other big cities as well. Why did Los Angeles and other big cities decide to set up pension systems for firemen and policemen in the late nineteenth and early twentieth centuries? How did the uniformed forces overcome the longstanding and deep-seated opposition to pensions for public employees? Also, why did the

LAFPPS's costs go up so much? Why did the number of retired officers and survivors rise so sharply, not only in absolute terms but also relative to the number of active officers? Why, in other words, did Los Angeles end up supporting two fire departments and two police departments? Next, why did the system's revenues lag so far behind its expenses? Why were the municipal authorities reluctant to raise members' contributions, increase the city's appropriations, and take other measures to build up the system's assets and hold down its mounting deficits? Finally, why were reformers unable to change the LAFPPS in ways that would have eased its financial predicament? Why did they fail in one effort after another to tighten eligibility requirements and reduce retirement allowances? What were the constraints on pension reform?

With the help of several Los Angeles officials—notably, M. Lewis Thompson, manager of the LAFPPS, and C. Erwin Piper, city administrative officer—I found the answers to these questions. I spelled them out in a paper entitled, "The Morass: An Essay on the Public Employee Pension Problem." The paper was presented at the Russell Sage conference and, along with the other papers, was subsequently published in *Social History and Social Policy.* When I finished the essay, I had no intention of doing more work on the firemen's and policemen's pension problem. But at the urging of several colleagues who were intrigued by my attempt to use social history to analyze a current problem, I decided to write a book on the subject. Instead of focusing just on Los Angeles, however, I opted to look also at New York, Detroit, San Francisco, Washington, D.C., and other big cities where the pension problem was extremely serious. I started the research on a grant from the Center for Studies of Metropolitan Problems of the National Institute of Mental Health and continued it on a sabbatical from MIT. I then received support from the Twentieth Century Fund, which enabled me to complete the research and write the manuscript. The result is *Pensions: The Hidden Costs of Public Safety*, a study of the firemen's and policemen's pension problem that is now plaguing most of the nation's big cities.

Acknowledgments

I have incurred a host of obligations in writing this book. The principal one is to the Twentieth Century Fund, which provided support for studying the pension problem and arranged for publication of the manuscript. For their help I would like to thank three members of the Fund's staff, James A. Smith, the late Walter Klein, and my editor, Pamela Gilfond.

Some of the research was done as part of my study of the Los Angeles Fire and Police Pension System, which was prepared for the Russell Sage Foundation Conference on Social History and Social Policy. My thanks go to Hugh F. Cline, the foundation's former president, as well as to David J. Rothman and Stanton Wheeler, who organized the conference. Some of the research was undertaken as part of a project on municipal employee unions that was housed at the Joint Center for Urban Studies of Harvard University and MIT and underwritten by the Center for Studies of Metropolitan Problems of the National Institute of Mental Health. For the Metro Center's assistance I would like to thank its head, Elliot Liebow. Lastly, I did some of the research while on sabbatical from

MIT. My thanks to MIT (as well as to the Department of Urban Studies and Planning, which has provided a congenial environment to work in for the past fifteen years).

A good many people went to great lengths to help me find the materials on which this book is based. Besides M. Lewis Thompson and C. Erwin Piper, I am particularly grateful to Jonathan Schwartz of the New York City Employees' Retirement System; Richard W. Brune of the New Orleans Finance Department; Thomas J. O'Brien, formerly of the District of Columbia Office of Budget and Management Systems; Walter L. Johnson of the Oakland Retirement Systems and Kenneth W. Maul, the former manager of the systems; Edmond J. Walsh, formerly of the San Francisco City and County Employees' Retirement System; Robert P. Clohessy of the Portland Fire and Police Disability and Retirement Fund; Paul Soullier of the Detroit Auditor General's Office; Tracy W. Howard of the Denver Budget and Management Office; and Joseph G. Metz, formerly of the New York State Permanent Commission on Public Employee Pension and Retirement Systems.

I am grateful as well to the staffs of the Citizens Budget Commission (New York City), the Chicago Civic Federation, the Minnesota Legislative Commission on Pensions and Retirement, and the New Orleans Bureau of Governmental Research. Thanks go also to the staffs of the National Archives; the Library of Congress; Harvard's Widener, Littauer, and Law libraries; the University of California's Governmental Studies Library; the MIT libraries; the Los Angeles Municipal Records Center; the State of Washington Division of Archives and Records Management; the Milwaukee, Chicago, Los Angeles, and New York City municipal reference libraries; and the Boston, Baltimore, Milwaukee, Minneapolis, and New York City public libraries.

I have learned a good deal about pensions and retirement from other historians and social scientists, most of whom are referred to in the notes. But I would like to call special attention to John Tessler, upon whose master's thesis I relied heavily for my analysis of pension reform in New York City, and three former students, Judy

A. Levenson, Janet M. Corpus, and Ilene G. Greenberg, on whose research papers I based much of the discussion of pension reform in Detroit and Los Angeles.

Many friends and colleagues helped in other ways. Bernard J. Frieden, S. Donald Gonson, Gary T. Marx, Lisa R. Peattie, Lawrence E. Susskind, Lloyd L. Weinreb, and W. Andrew Achenbaum read and criticized an earlier version of the manuscript. Pauline Maier commented on the historical issues analyzed in chapter 1, and Lance Liebman offered advice on the legal issues raised in chapter 5. Joseph Ferreira, Jr., patiently explained how to derive life expectancies from mortality tables. Jonathan Schwartz and M. Lewis Thompson not only read the manuscript and suggested revisions but also answered no end of questions about the arcane world of the actuaries and pension managers. My research assistants, Cindy Horan, Sherry Tvedt Davis, and Emily J. Tourin, went about their work with enthusiasm. Ms. Davis also made the tables and figures. My secretary, Jackie LeBlanc, did a splendid job typing the manuscript. My thanks to all of them.

Introduction

"We've got into a deeper and deeper morass," M. Lewis Thompson, manager of the Los Angeles Fire and Police Pension System (LAFPPS), told a reporter in 1972, "and there's no cheap way out." Thompson knew what he was talking about. The LAFPPS, which covered the city's 10,187 firemen and policemen, had on its rolls 6,358 retired officers and survivors, roughly two beneficiaries for every three officers, to whom it paid $41 million in benefits. To meet the system's current expenses and to amortize its prior obligations, the city council appropriated $63 million—45 percent of the fire and police departments' payrolls and 11 percent of the city's total expenditures. With another $18 million going to the City Employees' Retirement System, which covered the city's nonuniformed employees, Los Angeles spent more on pensions than on anything else except schools and police. While the LAFPPS's revenues exceeded its expenditures by $46 million in 1972, the system's obligations grew by even more. And its unfunded liabilities—a deficit equal to the difference between the system's liabilities and its current assets—jumped from $888 million in 1971 to $1.1

billion in 1972. That came to roughly $80,000 for each fireman and policeman and about six times the fire and police payroll.[1]

Five years later, the LAFPPS was in even worse shape. The number of retired officers and survivors came to 7,712, roughly three beneficiaries for every four officers. And the pension payroll approached $80 million, nearly twice as much as in 1972. On the advice of the system's actuary, the City Council put $113 million into the LAFPPS—49 percent of the fire and police payrolls and 12 percent of its total expenditures. As the council poured another $45 million into the City Employees' Retirement System, Los Angeles spent twice as much on pensions as on highways and three times as much as on parks and recreation. Spurred by sizable increases in property taxes and investment earnings, the LAFPPS's revenues went up from $83 million in 1972 to $160 million in 1977 and exceeded its expenditures in these years by $356 million. Nonetheless, the system's unfunded liabilities soared from $1.1 billion to $1.9 billion, more than $190,000 for each uniformed officer and close to nine times the fire and police payrolls. The LAFPPS "poses a substantial threat to the long-term financial health of the city," City Administrative Officer C. Erwin Piper warned in late 1977.[2]

Other cities were in much the same bind as Los Angeles. By the late 1970s, St. Louis had nearly two beneficiaries for every three uniformed officers, and Portland had more than three for every four. There were more than four beneficiaries for every five officers in Detroit and Indianapolis, more than six for every seven in Washington, D.C., and New Orleans, and more than nine for every ten in San Francisco. Oakland had as many persons on the pension rolls as on the department rosters, which helped to provoke in 1976 what Mayor John Reading called "the most serious financial crisis in the city's history." Things were even worse in New York City, where retired officers and survivors outnumbered active officers.[3] Outside the Southwest, most big cities were supporting two fire departments and two police departments—one made up of active officers and another, almost as large and growing even faster, made up of retired officers and survivors.

For most of these cities there is no relief in sight, at least not in the near future. The firemen's and policemen's pension rolls have been steadily climbing for decades. From the early 1940s to the late 1970s, they more than doubled in Denver, more than tripled in Portland, and went up even faster in Los Angeles and San Francisco. Moreover, the pension rolls have been growing more rapidly than the active rolls. Detroit, with just over one beneficiary for every four officers in 1941, had more than four for every five in 1977. Minneapolis, with two beneficiaries for every three officers in 1941, had close to seven for every eight in 1977. The pension rolls will continue to go up in the years ahead—and quite sharply in some cities. A case in point is Washington, D.C., whose firemen's and policemen's pension system was denounced by Senator Thomas F. Eagleton in 1978 as "far and away the premier ripoff system" in the nation. According to U.S. Treasury Department estimates, the system will have as many retired officers and survivors as active officers by 1991. Beneficiaries will outnumber officers by 1.2 to 1 in 2000 and by 1.6 to 1 thirty years later.[4]

As the pension rolls have grown—and as the retirement allowances, fueled by wage hikes and cost-of-living adjustments, have gone up—the pension payrolls have soared. Between 1967 and 1977 they more than doubled in Chicago, more than tripled in New Orleans, and more than quadrupled in Oakland. In most cities, the pension payroll has increased faster than the active payroll, which has risen more rapidly than most other municipal expenditures. Los Angeles is a good example. For every dollar the city spent on fire and police salaries, it poured about 18 cents into firemen's and policemen's pensions in 1941 and nearly 38 cents in 1978. By the late 1970s, the pension payrolls were enormous. They reached $294 million in New York, $59 million in Washington, $53 million in Detroit, and more than $20 million in several other cities. To give an idea of the magnitude of these figures, Detroit paid its retired officers and survivors slightly more than it spent on parks and recreation and twice as much as it spent on sanitation.[5]

The financial burden has been very heavy in cities where firemen's and policemen's pension plans have been operating on a funded or reserve basis—a system whereby the authorities put aside enough money each year to cover the costs of pension benefits as they are incurred. New York City put $332 million—roughly 46 percent of payroll—into the firemen's and policemen's pension systems in 1977, which was almost as much as it spent on capital improvements. The situation was even worse in Oakland. It poured about $18 million, or 71 percent of payroll, into the firemen's and policemen's pension fund in 1977, which was more than it spent on fire protection. The same year, Detroit put $72 million into the firemen's and policemen's pension system, which came to 54 percent of payroll and 10 percent of the city's total expenditures. Things were even worse in San Francisco, where Dan Mattroce, head of the retirement system, warned in 1979 that the city "faces a real possibility of bankruptcy" unless the retirement system is reformed. Between 1972 and 1978, the city's contributions to the firemen's and policemen's pension funds jumped from about 28 to 68 percent of payroll.[6] A few years later, San Francisco put about as much into firemen's and policemen's pensions as it spent on fire and police salaries.

The financial burden has been less heavy in cities where the firemen's and policemen's pension plans have been running on a "pay-as-you-go" basis—a system whereby the authorities put in only enough money each year to pay the retirement allowances that come due that year. But it will become much heavier in the years ahead. According to a 1976 study, Portland's contributions will rise from 26 to 52 percent of payroll in thirty years. A more recent study estimated that, if the system were fully funded, Portland would have to shell out 101 percent of payroll. In New Orleans, the actuaries reported that the city's contributions would go up from about $8 million a year in the mid-1970s to more than $27 million a year by the mid-1980s, a report that prompted Mayor Moon Landrieu to observe that the pension funds "are eating the city alive." In even worse shape were the District of Columbia's firemen's and policemen's pension funds, into which the government put nearly $60 million, or about 59 percent of payroll, in 1978. The actuaries estimated that, if the system remained on a pay-as-you-go basis, the District would be spending more on pensions than on salaries by the early twenty-first century.[7]

Although the cities have poured in a great deal of money, most firemen's and policemen's pension systems have built up massive deficits. By the late 1970s, the unfunded liabilities came to hundreds of millions of dollars in most big cities and exceeded $1 billion in New York, Los Angeles, and Washington, D.C. They were seven times as large as the fire and police payrolls in Minneapolis, eight times as large in New Orleans, and sixteen times as large in Oakland. For each fireman and policeman, they added up to $120,000 in Detroit, $170,000 in Portland, and $200,000 in Washington, D.C. The unfunded liabilities came to roughly one-seventh of the assessed value in New Orleans, one-fifth in Minneapolis, and one-third in Oakland.[8] A "fiscal time bomb," in the words of the Advisory Commission on Intergovernmental Relations, the huge unfunded liabilities were the prime symbols of urban America's public employee pension problem.

The firemen's and policemen's pension funds are an integral, if small, part of the nation's public employee retirement system. According to the Pension Task Force of the House Committee on Education and Labor, which made a thorough survey in 1975, there were at least 6,600 state and local plans—some with more than half a million members and some with fewer than half a dozen. These plans covered close to 10.4 million active employees and more than 2.3 million retired employees and survivors, to whom they paid more than $7 billion in benefits. Their revenues added up to over $21 billion, of which $10 billion came from employers, $5 billion from employees, and $6 billion from investments. And while the state and local plans had more than $108 billion in assets, they had at least $100 billion in unfunded liabilities. There were also 68 federal retirement plans, which covered more than 5 million active employees—about 2.2 million uniformed and 2.8 million nonuniformed—and nearly 4.5 million retired employees and survivors, who received nearly $14 billion in benefits. Their revenues totaled almost $18 billion, of which $13 billion came from the government, $3 billion from employees, and $2 billion from investments. These plans had over $40 billion in assets, but they were overshadowed by unfunded liabilities of at least $243 billion, two-thirds of which was accounted for by the military retirement system, a system that operates on a pay-as-you-go basis.[9]

From a national perspective, the plight of the firemen's and policemen's pension systems is not a pressing problem. Although these systems made up nearly two-thirds of the state and local pension plans in 1975, they covered only 4 percent of the nonfederal public employees. The California Public Employees' Retirement System had more active members than all the firemen's and policemen's plans combined. Only 7 percent of retired state and local employees and survivors were on the firemen's and policemen's pension rolls, and they received only 12 percent of the pension payroll. Although the state and local governments put more than $1.2 billion into the firemen's and policemen's pension systems in 1975, they contributed more than twice as much to the teachers' pensions systems and about eight times as much to all the public employee retirement systems. And although the firemen's and policemen's pension plans had massive unfunded liabilities, they made up only a small part of the total deficits of the state and local retirement plans and were dwarfed by the federal government's unfunded liabilities.[10]

But from the perspective of the nation's cities—especially its big cities—the plight of the firemen's and policemen's pension systems is an extremely pressing problem. To begin with, uniformed officers make up a large portion of the work force in the cities—much larger than in the states, counties, and townships. They are outnumbered by the teachers, who are by far the largest group of municipal employees. But with a few exceptions—among them New York and Washington, D.C.—the cities do not provide pensions for teachers, most of whom are members of statewide retirement plans. Thus the uniformed officers make up an even larger portion of the work force that belongs to municipal retirement plans. To give a few examples, the firemen's and policemen's pension systems covered more than one of every four municipal employees in

Detroit, more than one of every three in Chicago, and more than one of every two in Pittsburgh.[11]

Moreover, firemen's and policemen's pension plans are in much worse shape than retirement systems for nonuniformed employees. Los Angeles is a case in point. The Los Angeles City Employees' Retirement System (LACERS)—an "extremely conservative" system, to quote Manager Horace J. Enser, that covers the bulk of the city's nonuniformed personnel—had in 1977 about 2.4 times as many active members as the Los Angeles Fire and Police Pension System. Yet the LAFPPS—a "gold-plated system," in the words of City Administrative Officer C. Erwin Piper—had roughly 1.4 times as many retired members and survivors as the LACERS. With retired firemen and policemen receiving, on average, nearly twice as much as retired nonuniformed employees, the LAFPPS paid out more than twice as much in benefits as the LACERS. It also consumed more than two and a half times as much in property taxes. The city put $11,000 into the LAFPPS for each fireman and policeman and $2,300 into the LACERS for each nonuniformed employee; for each dollar it poured into the pension systems, firemen and policemen contributed 13 cents and nonuniformed employees 23 cents. And although the LAFPPS had almost as much in assets as the LACERS, it had more than three times as large an unfunded liability.[12]

Things were much the same elsewhere. In 1977, the New Orleans firemen's and policemen's pension plans had 1.4 times as many retired members as the city's retirement system for nonuniformed employees, which had 2.9 times as many active members. The same year, the Detroit firemen's and policemen's pension plans paid 1.6 times as much in benefits as the city's retirement system for nonuniformed employees, which had 2.6 times as many active members. San Francisco put 72 percent of payroll into the policemen's pension fund in 1977, 65 percent into the firemen's pension fund, and only 18 percent into the pension fund for nonuniformed employees. And in 1976, Minneapolis's unfunded liabilities came to more than $140,000 for each fireman, almost $100,000 for each policeman, and less than $30,000 for each nonuniformed employee.[13] When it comes to the big cities, it is safe to say that the public employee pension problem is at heart a firemen's and policemen's pension problem.

The problem has aroused a good deal of concern in recent years. According to Congresswoman Martha W. Griffiths of Michigan, public employee pensions are "time bombs ticking away in every American city." And to Jackson Phillips, vice-president of Moody's Investors Services, they are "the darkest cloud hanging over the cities." Alfred Parker, executive director of the Tax Foundation, has compared public employee pensions to an iceberg. Dean Lund, executive secretary of the League of Minnesota Municipalities, has likened them to an albatross. And a *Boston Globe* reporter has labeled them a "tapeworm." The *Institutional Investor* has asked, "Will pension costs push America's cities over the brink?" The *Nation's Business* has said that pension costs may "break the cities." And the *Wall Street Journal* has warned that a "Day of Reckoning" is at hand.[14]

The authorities have not been unmindful of these warnings. The New York State Permanent Commission on Public Employee Pension and Retirement Systems has studied the problem. So have the Pennsylvania Department of Community Affairs and the Minnesota State Planning Agency. The U.S. Senate Committee on the District of Columbia has looked at the problem in the nation's capital, where, in Senator Eagleton's words, "the time bomb . . . ticks away at a faster pace than elsewhere."[15] The U.S. House Subcommittee on Labor Standards has held hearings on pension reform, and the President's Commission on Pension Policy has looked into the nation's retirement programs. At the urging of local business and civic groups, the authorities have taken steps to put some big-city firemen's and policemen's pension systems on a sound financial basis. Usually over the opposition of the uniformed forces, they have tightened eligibility requirements, lowered retirement benefits, raised contribution rates, and/or improved funding arrangements. In some cases, they have even closed local firemen's and policemen's pension systems to new recruits, who are required to join less liberal statewide retirement plans. In the long run, these reforms will no doubt reduce pension costs.

In the short run, however, the reforms will probably not have much of an impact, for in most cases they apply only to new entrants, not to current members, who have vested rights in their pension plans.[16] Moreover, despite these reforms, the firemen's and policemen's pension rolls will grow in the years ahead—particularly in cities that greatly expanded their uniformed forces in the 1960s. Spurred by wage hikes and, in cities that provide automatic cost-of-living adjustments, by severe inflation, retirement allowances will go up, too. As a result, firemen's and policemen's pension costs will climb to record levels. Already in a serious fiscal squeeze, many cities will be hard pressed to pay their pension bills. They cannot count on much help from the states, which, in the wake of California's Proposition 13 and other tax-limitation and tax-reduction measures, are under strong pressure to hold down expenses. Nor can they expect much assistance from the federal government, which has long been unwilling to bail out local pension funds and is unlikely to reverse its position under the Reagan administration.

From time to time, some cities have failed to meet their obligations to retired officers and survivors. As the Washington, D.C., firemen's and policemen's pension funds were depleted in 1910, the commissioners postponed and later reduced retirement allowances; and they prorated them, on and off, from 1911 to 1916. The Cleveland firemen's and policemen's pension plans were so hard strapped for funds in the mid-1940s that they paid retired members and survivors only ten or eleven months a year. And as late as the mid-1960s and early 1970s, Hamtramck, Michigan, and a few communities in Arkansas, Mississippi, and Oklahoma temporarily suspended payments to retired officers and survivors.[17] But few cities have this option anymore. By virtue of a host of constitutional amendments, state laws, and judicial decisions, most cities are now prohibited from stopping, or even reducing, payments to retired employees. Unless a city goes bankrupt—in which case it would probably be up to the courts to work out in what order (and to what degree) the bills should be paid—it can no longer renege on its commitments. Under any condition short of municipal bankruptcy, retired officers and survivors will receive their full allowance.

What makes the firemen's and policemen's pension problem so serious is not that the cities may fail to meet their obligations to retired officers and survivors but that they will have to go to great lengths to do so. In order to pay their pension bills, they will be forced not only to raise property taxes and tap other sources of revenue but also to defer capital improvements, put off regular maintenance, reduce the municipal work force, and otherwise cut back vital services. "It will cost the city so much money to pay pensions that it won't be able to take care of the garbage," observed William R. Groves, a New Orleans actuary who had long urged elected officials to put the city's firemen's and policemen's pension plans on a reserve basis. As Senator Eagleton, one of the leading critics of state and local retirement systems, remarked, the day is coming when "some cities may have to choose between paying their pensioners and meeting their payrolls."[18] Thus the cities can meet their obligations to retired officers and survivors, but only at the expense of residents, taxpayers, and public servants. Firemen's and policemen's pension costs will not drive cities to bankruptcy, but they will severely exacerbate several of their most pressing problems.

One of the cities' problems—their "single greatest problem," according to Congress's Joint Economic Committee—is the deterioration of physical facilities. A case in point is New York City. Its two massive water tunnels—one built in 1917 and the other in 1936—have never been inspected, much less repaired. A third tunnel, now under construction, has been stalled by budget cutbacks. One-third of the city's streets should be reconstructed, and one of every ten bridges should be rebuilt or replaced. According to an Urban Institute study, New York City "needs a significantly increased rate of investment in maintenance and replacement if serious problems are to be avoided in coming decades." Much the same is true of Boston, Cleveland, New Orleans, and the nation's other older cities. Having failed to abide by regular schedules of maintenance, repair, and replacement, they are now plagued by leaky water mains, crumbling sewer pipes, and dilapidated streets and bridges. With many cities forced to cut back on capital spending in the mid-1970s, the situation would be even worse were it not for federal grants-in-aid. Obliged to pour a great deal of money into firemen's and policemen's pension systems, many cities will be hard pressed to reverse what the Joint Economic Committee has called the "downward trend in capital expenditures," a trend that "could result in a structural crisis" in the near future.[19]

Another of the cities' problems is the inadequacy of public services. Caught in a severe fiscal squeeze—for which the soaring firemen's and policemen's pension costs were partly responsible—many cities were forced to make cuts in operating expenses in the 1970s. As Groves predicted, New Orleans reduced trash collection from three to two times a week; New York cut pickups from five to three times a week, and Atlanta from twice to once a week. In addition to shelving plans to build a library, a health center, and a score of playgrounds and swimming pools, Philadelphia shut down its general hospital. Another victim of budget cuts, New York City's famous parks system was, in the words of a *New York Times* reporter, "dirty, unkempt, vandalized." Accompanying these cutbacks was a sharp reduction in the municipal payroll, the full impact of which was mitigated by the Comprehensive Employment and Training Act, under which the federal government paid the salaries of many municipal employees.[20] Forced to spend vast sums on firemen's and policemen's pensions, many cities will find it very hard to maintain, much less to raise, the current level of public services, an extremely grim prospect for many lower- and working-class Americans who rely heavily on the public sector for essential services.

Yet another of the cities' problems is what is widely regarded as an alarming decline in public safety. For at least two decades, Americans have been troubled by the sharp rise in crime, what

Newsweek called in 1965 "a condition of epidemic criminality." More recently, they have also been disturbed by a striking increase in arson—an increase that has reached "epidemic proportions," as one federal official put it in 1979. Fear of arson "has become a way of life for millions of Americans," a *New York Times* reporter found, especially for lower- and working-class residents of inner-city neighborhoods. Fear of crime, which extends well into middle- and upper middle-class communities, is even more common. Despite these fears, many cities have had to hold down fire and police costs. Unable to cut pensions or lower salaries, Boston, Denver, and San Francisco have imposed a freeze on hiring. New York, Philadelphia, and Detroit have resorted to layoffs. Perhaps the hardest hit was Detroit, which cut its roughly 5,600-man police force by 400 in 1979 and by close to 700 a year later, a reduction of about one-fifth.[21] As long as firemen's and policemen's pension costs are so high, many cities will be unable to hire additional officers, without whom the uniformed services say they cannot upgrade public safety in urban America.

Still another of the cities' problems is unequal access to public employment. Through the 1960s, most big-city police forces were overwhelmingly white, a pattern that reflected recruiting practices, entrance requirements, and, in some cases, out and out discrimination. The big-city fire departments had a worse record. In the aftermath of the 1960s riots, however, several civil rights and legal services groups mounted a campaign to open the uniformed forces to blacks and other minorities. And on the orders of elected officials and federal judges, many fire and police departments tried for the first time to increase the number of minority officers. The effort made good headway in New York, Detroit, Washington, D.C., and a few other cities in the early and mid-1970s. But in the face of hiring freezes and layoffs, it stalled and even lost ground shortly after. Of the roughly 700 Detroit police officers laid off in 1980, close to three-quarters were black and nearly half were women. As a result of the layoffs, the number of black and female officers, which had risen from 18 to 39 percent in the late 1970s, fell to 25 percent in 1980.[22] Forced to meet their obligations to retired officers and survivors—almost all of whom are white—many cities will be unable to hire many minority firemen and policemen, no matter how well qualified.

It is hard to tell precisely how much these problems have been exacerbated by firemen's and policemen's pension costs. But if New York City spent the same amount on pensions for each uniformed officer that it spent for each nonuniformed employee, the city would save about $220 million a year. With these savings, it could put half again as much money into capital outlays. By the same reckoning, San Francisco would receive a windfall large enough to reduce property taxes by more than 15 percent. New Orleans would save enough to expand the uniformed forces by just under 30 percent. And Detroit would be able not only to rehire the nearly 1,100 policemen who were laid off in 1979 and 1980 but also to appoint more than 1,000 additional officers.[23]

At the same time that firemen's and policemen's pension costs are growing, many big cities are contracting. They are losing population—in a few cases at a rapid rate. Revenues are lagging behind expenditures; and in states where the voters have adopted tax-limitation or tax-reduction proposals, they cannot catch up. To make matters worse, many cities are stuck with aging physical plants that sorely need to be repaired or replaced. They are also under pressure to slash essential services, reduce municipal payrolls, and even lay off uniformed officers, who have hitherto been sacrosanct. If all this were not enough, the benefits of the firemen's and policemen's pensions go mainly to whites, many of whom no longer live in the cities for which they once worked. And to the degree that the cities are forced to cut personnel and services in order to pay pension bills, the costs fall largely on blacks and other minorities. At a time when elected officials are looking for ways to solve the firemen's and policemen's pension problem, it may be instructive to analyze how so many cities got into the morass, why they have found it so hard to get out, and what if any lessons can be learned from this sorry story.

CHAPTER ONE

The Two-Tiered Pension System

The origins of the public employee pension problem go back to the late nineteenth and early twentieth centuries, when most big cities began to provide pensions for firemen, policeman, and other public employees.

The provision of pensions was a major departure from tradition. Down through the Civil War, pensions were unheard of in either the private or the public sector. Except under extraordinary circumstances, neither the cities nor the states accepted responsibility for the well-being of employees who grew old or incapacitated in their service. And long after France and England had set up retirement systems for superannuated and disabled civil servants, the federal government gave pensions only to ex-military personnel—mainly army and navy officers and veterans of the War of 1812, Indian Wars, and Mexican-American War. Moreover, in the course of establishing pension systems for public employees, most big cities set up separate plans for uniformed and nonuniformed employees.[1] These plans differed so much—in terms of eligibility requirements, retirement benefits, contribution rates, and funding arrangements—that they formed what can be described as a two-tiered pension system. This system, which remains largely intact today, laid the groundwork for the public employee pension problem and goes a long way toward explaining why the crux of the problem is the firemen's and policemen's pension plans.

To understand the emergence of the two-tiered pension system, it is necessary to bear in mind that through the late nineteenth century most Americans were strongly opposed to public employee pensions. Indicative of their attitude was the response in Congress in 1882 to proposals to provide pensions for the Revenue-Marine Service, which enforced the maritime laws, and the Life-Saving Service, which helped ships in distress. A pension system is "anti-republican" and "anti-American," argued Representative Strother M. Stockslager of Indiana. An outgrowth of "the effete monarchies of the Old World," it creates a privileged caste, an aristocracy of officeholders, that is entirely out of place in the United States. Unlike soldiers and sailors, civilian employees are not entitled to pensions, insisted Senator John T. Morgan of Alabama; they enter the service not to defend the nation but to make a living. The Life-Saving Service may be hazardous, Senator Samuel B. Maxey of Texas conceded, but its members "go into this service with a full knowledge of all the risks and dangers." Fire fighting is hazardous, too, Morgan observed, but "who has ever thought here of introducing a bill to pension the firemen of the District of Columbia?" To give pensions to the Life-Saving Service, Morgan warned, would be to drive a wedge, "which after a while will inflict upon this Government a civil pension-list whose burdens the taxpayers of this country will be unable to stand up under."[2]

In the face of such strong opposition to public employee pensions, the firemen, policemen, and nonuniformed employees were forced to find another way to protect themselves and their families against the hazards of old age, disability, and death. Following a well-established American tradition, they formed voluntary relief associations, which were in effect mutual aid societies. Firemen led the way, setting up associations in New York City (1792), Baltimore (1834), Cleveland (1839), and Minneapolis (1868). Policemen did not lag far behind, forming associations of their own in New York (1840), Providence (1870), Washington, D.C. (1872), and Kansas City (1883). Teachers and other municipal employees followed suit. The benefits varied from one association to another. The Baltimore Firemen's Beneficial Society provided relief to a member who was injured in the line of duty and to the family of a member who lost his life in service. The Providence Police Association gave a daily allowance to members if they were unable to work and a lump sum to their widows if they died. The Kansas City Police Association covered the member's funeral expenses and paid his widow a lump sum. The New York City Teachers' Mutual Life Assurance Association provided death benefits; and the Philadelphia Teachers' Aid and Annuity Association gave allowances in the event of sickness, disability, and superannuation as well.[3]

For a while, the relief associations served the firemen and policemen fairly well. One reason is that few officers grew old in the service. For much of the nineteenth century, most firemen were not, strictly speaking, in the service; they were volunteers who earned their livings as merchants, artisans, and laborers rather than as fire fighters. Unlike the firemen, the policemen were on the payroll. But the turnover was very high. Most patrolmen got their jobs through ward bosses, who regarded these posts as patronage for the political machine and rewards for the party faithful. The bosses who had the officers hired could also have them fired. A patrolman who offended a boss was soon looking for another job; and a change of administration was often followed by a purge of the department.[4]

Another reason relief associations were fairly successful is that few officers were disabled or killed on the job. The volunteer firemen spent most of their time carousing, brawling, socializing, politicking, and carrying on in other ways that eventually led the upper middle and upper classes to mount a campaign to replace them with paid firemen. Even when they attempted to put out a fire, they rarely went into a burning building. The big-city policemen passed much of their time eating, drinking, and visiting with friends.

They seldom did anything more hazardous than license vice, regulate crime, get out the vote, and otherwise promote the interests of the ward bosses and precinct captains.[5]

As the years went by, however, a number of changes took place that increased the chances that firemen and policemen would either grow old, be disabled, or die on the job. Following Boston's lead in 1837, one city after another replaced its volunteer fire companies with paid fire departments; by the early 1890s, when New Orleans joined the fold, the changeover was complete.[6] Although the firemen and policemen did not get formal job security until the late nineteenth and early twentieth centuries, a small but growing number of uniformed officers began to spend the bulk of their lives in the service. Fire fighting and policing became more hazardous, too. Firemen had to deal with the blazes that nearly wiped out Chicago in 1871 and destroyed much of Boston's business district a year later, the sort of conflagrations that had been uncommon before except in New York City. Policemen had to cope with the Draft Riots of the 1860s and the industrial disputes of the 1870s and 1880s, which dwarfed in magnitude and ferocity the ethnic and racial clashes of the antebellum years. To make things worse, the uniformed services came under strong pressure from the upper middle and upper classes to take whatever risks necessary to maintain public safety in urban America.

As the *Boston Globe* pointed out in the aftermath of the Tremont Temple fire of 1879, which left several officers badly injured, the changes in fire fighting and policing put a severe fiscal strain on the relief associations. It was one thing to give short-term allowances to a few firemen and policemen and small lump-sum payments to a few widows. It was quite another to provide long-term annuities to many superannuated and disabled officers and their survivors. The associations' income—which came from members' dues and initiation fees, proceeds from annual balls and concerts, fines for infractions of department regulations, rewards for the recovery of stolen property, and contributions from businessmen, property owners, and insurance companies—was not only inadequate but also unreliable. Some associations folded under the strain. Others would probably have done so were it not that many fire and police departments kept aged and incapacitated officers on the rolls, assigning them to look for fires, patrol cemeteries, carry messages, and work at what were called "desk jobs" in Boston, "soft snaps" in Chicago, and "light capacity" in the District of Columbia.[7]

At about the same time that firemen and policemen started to realize the limitations of their relief associations, they began to think of themselves as akin to soldiers. Pointing out that they risked their lives every day—exposed, as the *Firemen's Journal* put it in 1881, "to infinite peril"—their spokesmen claimed that firemen and policemen served the public in peacetime in the same way that soldiers served it in wartime. The uniformed forces were a "standing army," said Philadelphia Police Chief James Stewart, Jr., and Boston Fire Commissioner J. R. Murphy in the 1880s. One was engaged in a "war on crime," the other in a "war against fire." From this position it was a short step to the conclusion that firemen and policemen deserved the same benefits as soldiers—the most valuable of which was retirement pay. The *Firemen's Journal* took this line in 1877: "As soldiers, who devote their lives to the service of their country, are pensioned by the government when they become disabled or incapacitated for duty, so the Firemen, who devote their lives to the preservation of the lives and property of their fellowmen, should be provided for by the communities which they serve, when, from disability or length of service they become incapacitated for active duty." The same held for policemen, the *Rochester Democrat and Chronicle* insisted in 1887. In return for faithful service, "our local soldiers . . . are entitled to maintenance in their old age, and to care when disabled in performing their duty."[8]

Following this reasoning, firemen and policemen launched a campaign for pensions in New York and other big cities in the second half of the nineteenth century. They formed the vanguard of the movement for public employee pensions, watched, but for the time being not imitated, by the nonuniformed employees, who could not claim to have much in common with military officers. At first firemen and policemen only pushed for pensions for officers who were disabled in the line of duty. While some elected officials objected to what became known as disability pensions on the grounds that the officers took their jobs voluntarily, fully aware of the risks involved, this proposal generated only moderate opposition.[9]

Shortly afterward, however, policemen and firemen also called for pensions for officers who grew old on the job. The proposals for what were later referred to as service pensions—which ordinarily called for half-pay for officers who had put in at least twenty years and/or reached the age of fifty, fifty-five, or sixty, depending on the city—produced a storm of opposition. Although the information on the struggle over firemen's and policemen's pensions is sparse, what little there is helps to explain why the big-city firemen and policemen got pretty much what they wanted in spite of longstanding opposition to civil service pensions in the United States.

Many Americans objected to firemen's and policemen's pensions on the grounds that they were "un-American." Representative William S. Holman of Indiana took strong exception to a bill that would have authorized the District of Columbia to set aside up to $30,000 a year for its policemen's pension fund. The fund, which gave a maximum of $50 a month if an officer was disabled or killed either in the line of duty or in any way after serving fifteen years, was established by Congress in 1885. But the fund's income was limited to $1 a month deducted from each officer's paycheck. By 1894, the fund was so depleted that the House Committee on the District of Columbia recommended that it be given an appropriation out of the District's tax revenues. Speaking out against the proposal, Holman stressed that the Founding Fathers had rebelled against the British in part because of their extravagant pension list and had even thought of including a ban on civil service pensions in the Constitution. While he favored liberal pensions for soldiers and sailors, he considered pensions for civil servants "a most unmanly and discreditable imitation of the ideas of monarchies of Europe, from which our fathers hoped they had forever emancipated our country." If started, Holman warned, a pension list "will grow and expand until in the progress of time it shall have impoverished the masses of our people for the benefit of the few."[10]

Many Americans held that firemen's and policemen's pensions were "socialistic." Alderman Thomas W. Flood of Boston took this position in 1892 in a debate over a measure to provide pensions for policemen who were sixty-five years old or who were incapacitated after twenty years of service. If an officer is disabled in the line of duty, Flood said, "the city certainly should take care of him and his family." But there is no reason it should pay an able-bodied man anywhere from $600 to $1,200 a year for the rest of his life "for [doing] nothing" just because he has reached the age of sixty-five. The officer may have served the city faithfully for twenty or thirty years, but he was well paid for his labor, Flood pointed out. Alderman Weston Lewis took the same position, but carried it a bit further. He agreed that the city should provide disability pensions for uniformed officers, which it was already doing. But he sharply opposed a proposal to permit the authorities to retire a fireman on half-pay after fifteen consecutive years of service regardless of age. The city should not encourage able-bodied men to retire, Lewis said, much less give them the means to do so. Drawing on the traditional work ethic, he argued that "if a man is well and strong, or reasonably so, even if he has reached the age of sixty or sixty-five, there is no reason why he should not work so long as he is able to do so."[11]

Many Americans viewed firemen's and policemen's pensions as "paternalistic." Ever since the founding of the nation, they pointed out, it has been up to the individual to make "provision for his old age." As Lewis insisted, "It is the duty of every man to forecast the future as far as he can do, so that when he arrives at the age of fifty or sixty he will have something laid by with which to support himself and his family." Surely this was not too much to expect of policemen, Flood added. Besides getting many items at a discount, they earn enough to raise a large family, to save a little money, and, if they are worried about the future well-being of their dependents, to buy life insurance. Many Americans feared that if the state gave pensions to public employees it would do more than just undermine individualism and self-reliance. It would also discourage thrift and encourage profligacy. Once the employees were no longer required to set aside money for their old age, they would be tempted to acquire luxuries rather than to build up savings. A pension would therefore weaken self-control, self-restraint, self-denial, and other traditional values that served to hold down what one observer described as "the clamor of sensual wants."[12]

The opposition derided claims that firemen and policemen were akin to soldiers, who served for patriotic motives, earned a modest salary, and thus could not be expected to provide for their old age. " 'Tis not so!" one irate citizen wrote in 1893, in the midst of a struggle over a proposed pension system for St. Louis firemen. "They are there for the money and the soft snap that is in it. . . . There are thousands of good men who would be glad to have their jobs at a smaller salary." "This sentimental humbug about bravery should be stopped," wrote another St. Louis resident. "Every man in the dep[artmen]t is there for the good money there is in it."[13]

The opposition conceded that firemen and policemen take grave risks. But as the House Committee on the District of Columbia pointed out in 1884, the officers are fully aware of these risks when they enter the service, which is voluntary and "much sought after." Nor are firemen and policemen the only public employees who take such risks, Lewis argued. Pointing to the frequent cave-ins that left many employees injured for life, he said, "A man at work in the sewers runs great danger." "If the City of Boston is getting to be a charity bureau," Flood remarked, "I believe there is no man more deserving of a pension than is the poor, unfortunate laborer who has to earn his pittance by hard toil out in the streets with his pick and shovel"—and who gets only half as much pay as the fireman and policeman.[14]

If the authorities gave in to the firemen and policemen, the opposition argued, it would not be long before other public employees started to press for pensions. Speaking against a bill to set up pensions for Rochester policemen, Alderman Wesley Mandeville warned the council, "If you pass this tonight, you may expect a petition from the firemen next Tuesday night, another from the school teachers the next Tuesday night, and one from the other city employees the following Tuesday." (The *Rochester Post-Express*, which also opposed the bill, noted with sarcasm, "If the policy of the measure be good, why not extend it? Why not provide for pensioning the school teachers, the firemen, the street cleaners and other employees of the city? Surely we cannot have too much of a good thing.") Convinced that policemen had no better claim to a pension than other public employees, Mandeville feared that the authorities would be unable to resist these demands, a concern shared by many Americans. As they saw it, the provision of firemen's or policemen's pensions was the first step toward the creation of a full-fledged civil service pension system—a system that would put an intolerable burden on the taxpayers.[15] If firemen and policemen cannot make ends meet and put aside money for the future, the opposition argued, the authorities should raise salaries, not give pensions.

Spokesmen for the firemen and policemen dismissed these objections. Ignoring the charges that pensions were "un-American," "socialistic," and "paternalistic," they hammered away at the point that firemen and policemen were much like soldiers and should be treated as such. According to the *Firemen's Journal*, "Disabled men in a Fire Department are like wounded soldiers in a hospital when the army is engaged in an active campaign—they cannot do duty, they cannot be deserted, and, as they must be cared for, they constitute a serious draw back to the efficiency of the active force." A pension system is vital, the *Journal* went on, because "in the continuous battle waged against fire, there is no place for non-combatants." A fireman's job, like a soldier's, is "full of risk," said Fire Chief John Lindsay of St. Louis in 1893. "He must grope his way through stifling heat and smoke, climb to dizzy heights on frail ladders, under tottering walls, over slippery roofs or slippery floors." "From the moment the alarm strikes and the ride to the fire begins, until the order is given to 'pack up,' " Providence Fire Chief George A. Steere wrote the same year, "his life is in danger." "A policeman's duties are even more trying than those of a soldier," Providence Police Chief Benjamin H. Child added, and "quite as hazardous." It is "the voluntary exposure of life and limb," the *Rochester Democrat and Chronicle* declared in 1887, that distinguishes the perilous work of the policeman from "the routine duties of the clerk and teacher."[16]

Firemen and policemen, their spokesmen argued, should not be expected to provide for themselves and their families when age or injury renders them unfit for service. For one thing, most officers had no savings. As Superintendent William M. Dye of the Washington, D.C., police pointed out in 1884, the rank and file earn so little that a man with a family "is able to lay by little, if anything, for a rainy day." Out of his meager salary the officer also had to pay for his uniform, which Child noted was fairly expensive and had to be replaced from time to time. Moreover, policing and fire fighting were so hazardous that most officers could not get life insurance. As Alderman Thomas F. Keenan told the Boston City Council in 1892, "A life insurance agent who would take a policy on the life of a policeman or a fireman would be considered crazy, and would not hold his place in the company for an hour." Furthermore, most officers could not find other employment after leaving the force. "After a man has served as a policeman for twenty or thirty years," Alderman Walter E. Howes of Boston said in 1877, "he is unfit for any other duty, except that of private watchman." After putting twenty or more years in the service, the firemen have "practically unfitted themselves for other vocations," observed Lindsay.[17]

Spokesmen for the firemen and policemen took strong exception to the position that an officer was entitled to a pension if he was disabled in the line of duty but not if he had served twenty or more years. A policeman's work is so demanding, Child argued, that "after twenty or twenty-five years of service he is unfitted for further patrol duty." An officer who does his job well "is pretty well worn after fifteen years of police service," Superintendent Richard Sylvester of Washington, D.C., told a congressional committee in 1910. "A man can be a bookkeeper until he is eighty years old if his sight holds," Howes pointed out, "but a man cannot be a good policeman after he is fifty-five years old." The same was true of a fireman. After twenty years of service, Lindsay observed, he "is physically unfitted to perform the extraordinary duties or endure the hardships required in the service." Even if his health is still good, he has reached what Lindsay described as "premature old age." "One who has been in the service may survive to green old age," said the *Firemen's Journal* in 1884, "but after fifteen or twenty years' service as an active fireman in any of the large cities, he is pretty well battered up and worn out."[18] Stressing that twenty or more years of active service was the functional equivalent of disability, spokesmen for the firemen and policemen insisted that a superannuated officer was as much entitled to a pension as a disabled officer.

Firemen and policemen, their spokesmen argued, should not be denied pensions on the grounds that other municipal employees would soon demand equal treatment. Firemen and policemen have little in common with nonuniformed employees; their work is not only more hazardous, it is also more arduous. The fireman's job is one of "almost incessant toil," said Lindsay. On call twenty-four hours a day, he races from the fire house into the bitter cold, fights the blaze, and returns "coated with ice," his hands "blistered and bleeding" and his face "scorched [and] swollen." "If he goes through [the job] without contracting a fatal disease, and escaping maiming or crippling, he is a fortunate man," the *Firemen's Journal* stated in 1884. "He may fill some niche in business circles, but the best part of his life, his health and his energy have been expended in the fire service." The same holds for the policeman. Working ten or twelve hours a day, six or seven days a week, he "is continually on duty," Child wrote. He braves the cold in winter, Sergeant Edward Curry of the Washington, D.C., police said in 1910, and in summer, "he is out on a hot pavement, tramping and trudging from one end of his beat to the other until his feet blister from the heat." By the time his tour is over, "He is nearly parboiled." Little wonder, Curry added, that after twenty years a policeman "is not fit for a position that amounts to anything, anyhow, anywhere."[19]

Firemen's and policemen's pensions, their advocates argued, were not so much "a matter of sentiment" as "a cold-blooded business proposition" that would greatly enhance the efficiency of the uniformed services. A pension system would help the department get rid of what William McAdoo, police commissioner of New York City in the early 1900s, called the "dead-wood." The "dead-wood" were superannuated and disabled officers who were kept on the payroll because the departments were reluctant to condemn to poverty men who had rendered long and faithful service. A pension system would also help the uniformed services attract qualified recruits, a Chicago police official told a state legislative committee in 1887. It would be "the most encouraging inducement that we could offer to the policeman or the fireman," Sylvester said to the House Committee on the District of Columbia. A pension system would improve the quality of policing and fire fighting, too. Secure in the knowledge that their families were protected in the event of death or disability—which might otherwise serve as what the *St. Louis Post-Dispatch* called an "irresistible deterrent" to risk taking—the uniformed officers would do their utmost to stop crime and conflagration.[20]

The debate over firemen's and policemen's pensions raged in one city after another in the late nineteenth and early twentieth centuries. But in the end, the authorities gave in to the uniformed officers. One reason was that the opposition's case was seriously undermined by changes taking place in America. The charge that public employee pensions were "un-American," "socialistic," and "paternalistic" lost a lot of credibility when the American Express Company, the Baltimore and Ohio Railroad, and several large corporations started to set up pension systems for their employees.[21]

The argument that the government should not treat firemen and policemen differently from nonuniformed employees was not compelling when the military analogy was emerging as the conventional wisdom of the upper middle and upper classes. The *Detroit Free Press* used this analogy to strengthen the case for firemen's and policemen's pensions in 1885. Speaking in favor of a bill to provide pensions for disabled and worn-out firemen being considered by the Michigan Legislature, the paper said, "It is not, perhaps, a logical demand. A stern application of business principles would probably rule it out and say that the fireman takes the risks of service voluntarily. But the same business principles would rule out also the claim of the soldiers, a claim which the public has always been ready to recognize and approve."[22]

The argument that a man should work as long as he is able to lost much of its force as many Americans came to believe in what W. Andrew Achenbaum has described as "the obsolescence of old age." Abandoning the traditional view of the elderly as a source of "moral wisdom and practical sagacity," they began to regard old age as a form of disability—in one scientist's words, "a chronic, infectious disease." Less impressed than their fathers by the intrinsic value of work, many Americans started to think that after thirty or forty years on the job a man should spend his remaining days in what soon became known as retirement. The last stage of life should be geared not to work but to leisure, which should not be denied a man just because he was still in good shape. Also severely weakened was the contention that the firemen and policemen should lay aside enough money while employed to take care of themselves and their dependents after they could no longer work. Many Americans were starting to doubt that thrift alone was adequate protection against the perils of death, disability, and superannuation. They were also beginning to lose confidence in the voluntary associations which dealt with the problems of old-age dependency. These concerns would later generate a nationwide movement to persuade the government to give old-age pensions to all citizens regardless of occupation.[23]

Another reason the authorities gave in to the uniformed officers was that they had the support of two powerful groups—both of which favored firemen's and policemen's pensions, although for different reasons. The first was the political machines, which governed most American cities in the second half of the nineteenth century and which saw in firemen's and policemen's pensions a way to increase turnover and thereby expand patronage. In 1878, for example, Tammany Hall, New York City's Democratic machine, supported a bill that empowered the police commissioners to retire an officer after twenty years of service as part of a scheme to get rid of two hundred veteran officers and replace them with party loyalists. In the 1890s, two state legislative committees revealed how the machines exploited the pension systems. The Lexow committee, which investigated the New York City police force, found that several officers, some in good health, able and willing to work, had been forced to retire "to make room for somebody else," as one captain put it, and to provide patronage for the police commissioners. The Berry committee, which probed the Chicago police department, reported that many officers with long experience, exemplary records, and good health "were removed and placed up on the pension roll, solely for the purpose of giving places to some other men [presumably party loyalists] who could not be promoted otherwise."[24]

The second group was the Progressives. Upper middle- and upper-class native Americans, drawn mainly from the commercial, financial, and professional elites, these reformers were out to destroy the political machines. They were also intent on upgrading the quality of policing, fire fighting, and other public services. Most Progressives viewed pensions as an integral feature of a modern personnel system. A pension system would help remove many disabled and superannuated firemen and policemen who could no longer carry their weight and, as McAdoo claimed, were "bitterly opposed to innovations however beneficial." Their departure would enable younger, more vigorous men to join the forces and would permit capable officers to move up the ranks.[25]

The Progressives also believed that a pension system would attract many qualified recruits who might not otherwise have considered a career in policing or fire fighting. Provided that the retirement benefits were not vested—provided, in other words, that unless disabled an officer could not collect a pension until he put in at least twenty years on the force and/or served until fifty-five or sixty—a pension system would discourage veteran officers from leaving in their most productive years.[26] A pension system might even deter some policemen and firemen from taking payoffs, abusing citizens, and otherwise violating department regulations because if caught they stood to lose not only their jobs, which was bad enough, but also their pensions, which for many veterans was even worse.

Yet another reason the authorities gave in to the uniformed officers was that most Americans did not think pensions would cost much. Advocates argued that the systems would be pretty much self-sustaining; a few implied that they might produce a surplus. The opposition sharply disagreed. But few Americans knew how to evaluate the conflicting claims, and, with very little data as yet available, most tended to underestimate pension costs. Writing about a proposed pension plan for policemen, the *Rochester Democrat and Chronicle* declared in 1887 that it would cost the taxpayers only $1,000 a year, with the bulk of the revenue coming from deductions from officers' salaries and fines for infractions of department rules. A special committee of the Providence City Council reported in 1892 that a firemen's and policemen's pension system "would not add any perceptible burden upon the taxpayers [who] would cheerfully assume so slight an increase, if any, in the amount required to be raised by taxation." And in 1913, by which time a good deal of evidence to the contrary was available, the *Oregon Daily Journal* contended that a proposed pension system for Portland firemen would cost the average taxpayer around "ten to fifteen cents a year for not more than five years."[27] At this price, the demands for pensions were hard to resist.

New York was the first city to accede to these demands. It gave rudimentary disability pensions to policemen in 1857 and to firemen in 1866. Full-fledged service pensions were granted to policemen in 1878 and to firemen in 1894. Many big cities—Chicago, Boston, Detroit, and San Francisco among them—followed New York's lead in the 1880s and 1890s. Others, including Los Angeles, New Orleans, Atlanta, and Minneapolis, did not lag far behind. A few bucked the trend, but these were the exceptions, and in time they came around. In a survey of old-age dependency in the United States, Lee Welling Squier wrote in 1912 that "the impression seems to be widespread that, as it is incumbent upon the nation to provide pensions for the soldiers and sailors who hazard their lives in the nation's defense, so it is incumbent upon the cities to make as liberal provision for the citizens who hazard their lives in the protection of the life and property of their fellow-citizens." Squier went on to predict—quite correctly—that "it is only a question of a short time until every city of any considerable size and importance throughout the land will make provision for its defenders of life and property who have been worn out or disabled in the service."[28]

As the opponents of firemen's and policemen's pensions predicted, other municipal employees soon began to push for pensions. In the forefront were the teachers. They made initial efforts in New York City in the 1880s and then launched vigorous campaigns in the 1890s, not only in New York but also in Chicago, Detroit, and Cincinnati. Other nonuniformed employees followed suit. Boston's laborers asked for pensions in the 1890s, as did the employees of New York City's health, finance, and street-cleaning departments and the faculty of the College of the City of New York.[29]

By the first decade of the twentieth century, the campaign for public employee pensions was well under way. Its spokesmen argued that nonuniformed employees were entitled to the same benefits as uniformed officers. As a representative of the Federation of State, County and Town Employees told the Massachusetts Commission on Old Age Pensions, Annuities and Insurance in 1908, "We believe that the state should give us the same pensions that it now gives the firemen and the policemen." Pointing to the mechanics and laborers who worked on the tunnels, bridges, and street lights, the spokesmen contended that their jobs were no less hazardous, arduous, or vital than the firemen's and policemen's. A few insisted that nonuniformed employees had a stronger claim to pensions than uniformed officers because their salaries were much lower.[30]

The campaign for public employee pensions ran into several of the same objections as the campaign for firemen's and policemen's pensions. If an employee knew he had a pension coming, a former state employee said to the Massachusetts Commission on Old Age Pensions in 1908, he would have no incentive to provide for himself and his dependents in the future; his independence and self-esteem would be eroded. A pension system, another opponent argued ten years later, would put "a premium on thriftlessness." Public employees enter the service voluntarily, the opposition insisted; they earn enough to protect themselves and their families against the problems of disability and superannuation. A pension system, a Fall River resident told the commission, would weaken the family, because children would no longer feel responsible for the future well-being of their parents. It would also undermine the longstanding tradition of voluntary relief. Public employees were no more entitled to pensions than private workers, the opposition stressed; it would be unfair to favor the public sector and pass the costs on to the private sector. Worst of all, said one member of the Massachusetts Constitutional Convention of 1918, public employee pensions would be "the entering wedge for a general system of disability or old age pensions," a wedge that would open "the floodgates of State socialism."[31]

Spokesmen for the public employees responded that nonuniformed employees were entitled to the same benefits as uniformed officers. They argued that with wages so low and costs so high, most municipal employees could barely make ends meet; and as State Representative Cornelius J. Carmody told the Massachusetts Commission on Old Age Pensions, "It is impossible for them [to] save any money." To set disabled and elderly employees adrift without means of support would be inhumane; to expect their children to take care of them would be unduly optimistic; and to send them to the poorhouse would not be much of a reward for long and faithful service. If the community was ready to support a public employee in the poorhouse, one minister asked, why should it object to supporting him on a pension? As far as charges of state socialism went, spokesmen for the nonuniformed employees pointed out that many railroad companies and other large corporations were already providing pensions for their employees. Anyone who thought the costs of pensions would be excessive, Boston Street Superintendent Guy C. Emerson remarked, should consider the expense of maintaining hundreds of disabled and superannuated employees on the municipal payroll.[32]

With the exception of the teachers, who prevailed on the authorities to set up pension systems in several cities at the turn of the century, the nonuniformed employees made little progress. Many Americans still viewed public pensions as un-American and anti-republican—a step toward the establishment of a privileged class of officeholders. They approved of pensions for soldiers and sailors and, albeit reluctantly, for firemen and policemen, but not for other public employees. Thus by the early twentieth century, a time when most European cities provided pensions to all their employees, New York was the only American city to give pensions to any of its employees aside from policemen and firemen. And it gave them only to members of the health, finance, and street-cleaning departments. As F. Spencer Baldwin, a strong advocate of public pensions, observed in 1911, "No American city has yet established a general plan of superannuation allowances for its employees."[33]

The campaign for public employee pensions picked up considerably in the second decade of the century, when the Progressives gave their support to it. The Progressives were profoundly distressed by the inadequacy and inefficiency of public service in American cities. Setting out to do something about these problems, they discovered that most cities had men of sixty-five, seventy, and even eighty on the payroll. The Progressives were not without sympathy for department heads who were reluctant to dismiss employees with no other means of support after long years of faithful service. But believing that old age was synonymous with obsolescence, the Progressives held that this practice was extremely unproductive. They argued that superannuated employees not only lowered the quality of public service but also raised the costs. Apart from keeping out qualified recruits, they blocked the promotion of younger employees, destroying their initiative and undermining their morale. As long as department heads had no option other than to remove or retain superannuated employees, the Progressives concluded that it would be impossible to upgrade the public service.[34] In the interest of reform, the cities would be better advised to set up retirement systems than to continue to provide what were in effect hidden pensions.

A well-designed pension system, wrote Baldwin not long after his stint as executive secretary of the Massachusetts Commission on Old Age Pensions, is "a plain business proposition," one that would "promote a higher degree of efficiency and economy in the municipal service." How it would do so was spelled out by Lewis Meriam in *Principles Governing the Retirement of Public Employees*, a study published in 1918 by the Institute of Government Research (now the Brookings Institution). A pension system, Meriam pointed out, would do more than just enable the government to get rid of many aged workers who could no longer carry their weight, thereby raising the efficiency and lowering the costs of services. It would also upgrade the personnel by attracting qualified applicants who would otherwise go into the private sector, reduce turnover by providing an incentive to stay, and enhance morale by increasing opportunities for advancement. As the secretary of the Indiana Board of Charities said in 1906, a pension system might even reduce misconduct because many employees would be reluctant to do anything that might jeopardize their retirement benefits. Replying to critics who objected to public employee pensions on the grounds of cost, Baldwin spoke for most Progressives when he argued that a well-designed system "would, in the long run, diminish rather than increase the taxpayers' burden."[35]

During the 1910s, however, several studies alerted the Progressives to the pitfalls of public employee pensions. The New York Bureau of Municipal Research, a prominent reform group, reported in 1913 that the city's policemen's pension system had run up liabilities of more than $65 million and that when it matured the pension payroll would come to roughly one-third of the active payroll. A few years later, the New York City Commission on Pensions disclosed that the firemen's pension system was even worse off. In 1917, the Illinois Pension Laws Commission announced that the state's pension funds were insolvent and that in Chicago alone the deficits came to more than $30 million for policemen, more than $13 million for firemen, and more than $5 million for teachers. Other studies revealed that the firemen's and policemen's systems were in serious trouble in Buffalo and Washington, D.C., and that a few teachers' pension systems had already collapsed. Combined with the New York City Commission on Pensions' findings that in some Western European cities pension payrolls came to anywhere from 17 to 37 percent of active payrolls, these revelations were extremely disquieting.[36]

After these studies were released, the Progressives modified their position in a way that set them apart from the employee groups, which naturally wanted as liberal pension plans as possible. The reformers would campaign for pensions only if the proposed plans were "sound." By "sound," Baldwin, Meriam, Paul Studensky of the Bureau of Municipal Research, George B. Buck, probably the foremost public sector actuary of the time, and other Progressives meant several things. First, the proposed pension systems should make clear what the benefits would cost in the long run and where the money to pay for them would come from. Second, the systems should be contributory. Since a pension plan would serve the interests of both employee and employer, providing security for one and increasing efficiency for the other, each should bear a fair share of the costs, preferably one-half. Third, the systems should be funded. The authorities should put aside enough money to pay for the benefits as they accrued, not when they came due. Each generation of taxpayers should pay its pension bills rather than pass them on to the next generation. Fourth, since pensions are in part a reward to the employees for long and faithful service, the retirement allowances should be pegged to years of employment, not simply tied to final salary.[37]

Led by employee associations and reform groups, the campaign for public pensions gained a good deal of momentum after 1910. In some cases, elected officials took proposed amendments to city charters drafted by employee groups and submitted them to the voters. But in many cases, they set up special commissions to look into the issue and to come up with recommendations. Among the cities, New York led the way in 1916, with Milwaukee, Baltimore, and Oakland following suit not long after. Massachusetts set the pattern for the states in 1910, with Illinois, New York, New Jersey, and Pennsylvania not far behind.[38]

The commissions relied heavily on the Progressives and the actuaries. Baltimore's Retirement Commission followed the recommendations of the city's Commission on Efficiency and Economy, an elite group that asked for help from the Baltimore Bureau of Governmental Research and from Buck, who had done much of the work for the New York City Commission on Pensions. The Oakland Civil Service Board sought the advice of William Leslie, an actuary who had worked for the San Francisco retirement fund, and of the San Francisco Bureau of Management Research, one of the Bay Area's many reform organizations.[39] Hence it was not surprising that the commissions usually came up with proposals that bore a close resemblance to what the Progressives defined as a sound pension system.

In some cities, the proposed pension systems were put into effect without much of a struggle. Philadelphia set up a pension plan for its nonuniformed employees in 1915, less than a year after the Pennsylvania legislature passed an act requiring it to do so. Five years later, New York City followed Philadelphia's lead. Inspired by a report of the Illinois Pension Laws Commission, which in 1919 drafted guidelines for state pension policy and incorporated them into legislative proposals, the Illinois legislature established a pension system for Chicago's nonuniformed employees in 1922. After doing a thorough study, the Baltimore Commission on Efficiency and Economy came out in 1923 in favor of public pensions and recommended the formation of a special commission to work out the provisions. The commission, which consisted of officeholders and prominent citizens, presented a retirement plan to the city in 1925; and the City Council adopted it a year later. After rejecting one charter amendment in 1915 that would have set up a pension plan for San Francisco's nonuniformed employees, the voters approved another five years later—a more restrictive amendment that had the backing of most business and civic groups. A board of administration, appointed by the mayor, drew up the plan, which was adopted by the San Francisco Board of Supervisors in 1922.[40]

In other cities, the campaign for public employee pensions ran into serious trouble. In the face of strong opposition by the Boston Finance Commission, the Boston City Council declined in 1911 to accept a state act that would have enabled it to provide pensions for nonuniformed employees. Not until 1922, thirty years after the council first debated the issue, were these employees covered. Seattle voters twice rejected a proposed pension system for nonuniformed employees in the 1910s. Not until 1927, fourteen years after the campaign was launched, did the voters approve a charter amendment on the basis of which a pension system was established in 1929. The Milwaukee City Council refused to go along with the recommendations of the Milwaukee Pension Laws Commission, which drafted in 1920 what many regarded as a model pension system. On the eve of the Great Depression, the city's nonuniformed employees still lacked a retirement plan. The same was true in Los Angeles, where, in spite of vigorous efforts by the All City Employees' Association and occasional support by reform groups, the electorate turned down proposed charter amendments—in 1918, 1920, and 1926—that would have provided pensions for nonuniformed employees.[41]

The campaign ran into serious trouble for several reasons—aside from the longstanding opposition to public employee pensions. By the early 1920s, many Americans who would have favored pensions for nonuniformed employees were so dismayed by the rising costs of firemen's and policemen's pensions that they were reluctant to take on any additional obligations. Pointing to the $1.5 million deficit of the Los Angeles firemen's and policemen's systems, the Municipal League stressed in 1922 that it would not back any pension plan for nonuniformed employees until an actuary made a study of the costs. Moreover, policemen and firemen, intent on preserving their own pension systems, sometimes opposed attempts to set up comprehensive retirement plans covering both uniformed and nonuniformed personnel. According to the Carnegie Foundation for the Advancement of Teaching, a strong advocate of public pensions, policemen helped to defeat state legislation in the early 1920s that would have given pensions to nonuniformed employees in Boston and Providence. Finally, nonuniformed employees themselves sometimes objected to a proposed pension system because they were displeased with the plan's provisions. Seattle electrical workers opposed a plan in 1914 on the grounds that the age requirement was too high; and Los Angeles municipal employees fought a plan in 1926 on the grounds that the retirement allowances were too low and the contribution rates too high.[42]

The campaign ran into other obstacles as well. Reform groups would only give their support to what they regarded as sound pension systems. In 1910, the Boston Finance Commission urged the City Council to reject a state enabling act largely because it did not provide for compulsory retirement of superannuated workers. The Milwaukee Citizens' Bureau objected to the proposals of the Milwaukee Pension Laws Commission in 1921 because the city would be required to contribute twice as much as the employees. The Portland City Club opposed a pension plan in 1930 on the grounds that the age requirement was too low, the retirement allowance was too liberal, and the taxpayers would have to assume the entire deficit of the proposed system. To make matters worse, the Depression left many cities destitute. At a time when elected officials were hard pressed to meet payrolls and property owners were hard put to pay taxes, the politicians found it difficult to justify spending additional funds for the benefit of anyone who was fortunate enough to be gainfully employed. The All City Employees' Association of Los Angeles incorporated a pension plan into a charter amendment in 1930, but decided against submitting it to the voters because economic conditions were so bad. The Oakland City Council was so strapped for funds in the early 1930s that it could not start up a retirement system for nonuniformed employees that was authorized by the voters in 1927.[43]

Despite several setbacks, the campaign for public employee pensions made a good deal of headway. By the late 1920s, roughly half of the nation's twenty largest cities—among them New York, Chicago, Philadelphia, and Detroit—were providing pensions for nonuniformed employees, as were Massachusetts, New York, Pennsylvania, and a handful of other states. Spokesmen for public employees put the remaining cities on the defensive, charging them with failing to keep abreast of modern trends and pressing them to follow the example of more progressive communities. Moreover, in 1920, after close to thirty years of pressure from employee organizations and reform groups, Congress established a retirement system for federal civil servants, an act that dealt a heavy blow to the view that public pensions were un-American and antirepublican. At the behest of the Roosevelt administration, Congress also passed the Social Security Act of 1935, the culmination of decades of struggle for old-age pensions. The act covered most private workers, many of whom were by then members of corporate pension plans. But it excluded all public employees, who it was assumed would be protected by retirement systems of their own, and thus severely undercut the position that the public sector should be treated no better than the private sector.[44]

Ironically, the Depression, which at first held back the campaign for public employee pensions, later helped it along. Advocates had predicted that in the absence of a retirement system many superannuated and disabled employees would hang on to their jobs as long as possible. As the economy collapsed, this prediction was borne out. In 1935, the Milwaukee Civil Service Commission reported that one of every six city employees was over sixty and one of every eleven over sixty-five. The low turnover not only raised costs and lowered efficiency, the commission contended, but also deprived young people of access to municipal employment.[45]

A different problem arose in Los Angeles, where the California Supreme Court held in 1935 that the city could not attempt to cut costs by putting employees on half time. The Los Angeles parks commission thereupon laid off fifty people, some of whom had served for thirty or forty years. After such long service, a layoff would have been harsh under any conditions, but it seemed inhumane at a time when the private sector had few openings and the relief system was coming apart. To many residents of Los Angeles, it appeared that the only way to resolve the dilemma was to establish a retirement system for the city's nonuniformed employees.[46]

By the mid-1930s, the pressure for public employee pensions was irresistible. At the urging of the Civil Service Employees' Association, the Oakland City Council appointed a board of administration to draft a pension system for nonuniformed employees in 1934. Three years later—ten years after the voters authorized the system and twenty years after the employees launched their campaign—the council finally set up the plan. At the request of the Milwaukee Civil Service Commission, the Milwaukee City Council formed a special committee in 1936 that came out in favor of pensions for nonuniformed employees. A year later, the City Council unanimously approved a plan, the details of which had been worked out by Buck and incorporated into an enabling act passed by the Wisconsin legislature. In Los Angeles, where the City Council had authorized a pension system for nonuniformed employees in 1935 but failed to appropriate funds for it, the All City Employees' Association drafted another plan, which was put on the ballot in 1937. The voters, who had turned down similar proposals before, adopted the measure by a large majority.[47] With a few exceptions, all the big cities provided pensions for nonuniformed employees by the late 1930s—a testimony to the political influence of the employee associations and reform groups.

To have overcome the deep-seated and longstanding opposition to civil service pensions was quite a feat for the nonuniformed employees, a feat in which the employee associations and reform groups took great pride. But if the nonuniformed employees had done well, they had not done nearly as well as the firemen and policemen, whose pension systems were much more liberal. While the firemen's and policemen's pension plans varied somewhat from one city to the next, as did the plans for nonuniformed employees, these variations were dwarfed by the striking differences between the plans for uniformed and for nonuniformed employees. On the basis of two surveys—a study of firemen's and policemen's pension plans done in 1909 and a study of nonuniformed employee plans made in 1927—it is possible to spell out the salient differences.

Consider eligibility requirements. When it came to disability pensions, the firemen's and policemen's plans did not differ too much from the plans for nonuniformed employees. In both cases, most plans provided pensions for personnel disabled in the line of duty, regardless of age and length of service, although firemen and policemen were more likely to be disabled on duty than nonuniformed employees. Many plans also gave non-service-connected disability pensions (which tended to be smaller than service-connected disability pensions), but as a rule only to personnel who had put in from five to twenty years of service. Most plans paid benefits as well to the dependents of an employee who died in the line of duty or who succumbed to natural causes after long service. When it came to service pensions, however, the firemen's and policemen's plans differed very much from the plans for nonuniformed employees. To be eligible for a service pension, nonuniformed employees typically had to reach sixty—and, if the retirement allowance was to amount to much, put in twenty to thirty years on the job. While firemen and policemen were commonly required to serve twenty or twenty-five years before retirement, in some cities the age requirement was only fifty or fifty-five, and in many others uniformed officers could retire after twenty to twenty-five years of service at any age.[48]

There were also striking differences in retirement benefits. Most firemen's and policemen's pension plans pegged the retired officer's allowance to his salary—a policy that was probably modeled on the retirement system for military officers. Half-pay was customary, although in Boston and New Orleans a totally disabled fireman could get as much as two-thirds. For most officers, the allowance was based on salary at time of retirement. With a couple of exceptions, few big cities took years of service into account in computing benefits for uniformed personnel. Some retirement plans for nonuniformed employees followed the practice of the firemen's and policemen's pension systems, but, guided by Progressive principles, most plans for nonuniformed employees pegged benefits to years of service and/or member contributions. The Baltimore Employees' Retirement System was typical. At the age of sixty an employee was entitled to an annuity, the amount of which was determined by the value of his accumulated contributions, plus a pension of equal value, which was paid for by the city. Under the plan, an employee would ordinarily have to work for thirty-five years in order to retire at half-pay.[49]

Contribution rates varied as well. Most firemen's and policemen's plans were wholly or largely noncontributory. Firemen put in nothing in Baltimore, 1 percent of salary in Chicago, and 2 percent in Los Angeles. Policemen contributed nothing in Buffalo, 1 percent of salary in Denver, and 2 percent in New York. Not even in Milwaukee, where the firemen and policemen put in 2.5 percent, did the officers' contributions bear a relationship to their benefits, much less meet half the system's costs, the share recommended by the Progressives. By contrast, most nonuniformed employee plans were contributory. Employees shelled out 2.5 percent of salary in Pittsburgh, 4 percent in Boston, and roughly 4.5 percent in Chicago. In several cities that strictly followed Progressive principles, the contribution rates were tied to entry age and other variables. Under these arrangements, the contribution rates ranged from 2.9 to 6.4 percent in San Francisco, from 3 to 8 percent in Minneapolis, and from 3.8 to 8.3 percent in New York City.[50]

There were also differences in funding arrangements. Virtually all the firemen's and policemen's plans operated on a pay-as-you-go basis. Instead of setting aside adequate funds to cover pension costs as they were incurred, they just put in enough money to pay retirement allowances as they came due. The principal sources of revenue were fees, fines, and, in some cities, member contributions and investment earnings. If these sources fell short, the plans appealed to the cities (and sometimes to the states) for additional funds, which normally came out of general revenues. None of the big cities attempted to put aside enough money to meet long-term costs—which, in any case, had not been calculated—much less to amortize prior obligations. A few plans for nonuniformed employees operated in a similar way. But in most big cities they ran on a reserve basis, abiding by the Progressive principle that each generation of taxpayers should pay its pension bills rather than pass them on to the next generation. The cities contributed a sum—usually expressed as a percentage of payroll—which, along with member contributions and investment earnings, was supposed to build a fund adequate to meet the system's current and future expenses. Several cities put in additional funds, also expressed as a percentage of payroll, to amortize the prior service liabilities as well.[51]

The two-tiered pension system was largely a product of the chronology of the movement for public employee pensions. The firemen and policemen were very much in the vanguard of this movement. Of the few models available to them, by far the most significant was the federal government's retirement plan for military officers. This plan, set up at the start of the Civil War, provided three-quarters of final pay to commissioned officers who had served for thirty years or who were disabled in the line of duty. Entirely noncontributory, it ran on a pay-as-you-go basis, with its expenses treated as part of the military budget.[52] In their enthusiasm for the military analogy, firemen and policemen incorporated many of the cardinal features of this plan into their proposed pension systems.

The elected officials to whom the proposals were submitted were impressed by the dangers of policing and fire fighting, not to mention the political clout of the uniformed forces. Most of them did not have strong views about how long firemen and policemen should work or how much they should get at retirement. Nor did they have a clear idea of the long-run costs or how the costs should be apportioned between employee and employer and between one generation and the next. Thus elected officials approved proposed pension plans without looking too closely at their underlying principles.

By the time nonuniformed employees began to push for pensions of their own, things were quite different. From the grim reports that were issued in the 1910s, many Americans realized that firemen's and policemen's pensions were much more expensive than they had been led to believe. They also knew a good deal about the long-term costs and unfunded liabilities of the existing systems. The Progressives had also come up with what they regarded as the principles of a sound pension system, according to which the plans for nonuniformed employees should have tighter eligibility requirements, lower retirement benefits, and higher contribution rates than the firemen's and policemen's systems. These plans should operate on a reserve, as opposed to a pay-as-you-go, basis, too. The support of the Progressives was critical because on their own the nonuniformed employees would probably have been unable to overcome the opposition to civil service pensions. They were less well organized than the firemen and policemen, had less political influence, and could not claim to be quasi-military personnel. Not only did the nonuniformed employees have to draft pension plans that met the criteria of the Progressives and passed the scrutiny of the actuaries, but, to win the approval of the authorities, they were sometimes forced to tighten these plans in ways that set them even further apart from the more liberal firemen's and policemen's pension systems.

From a fiscal standpoint, the retirement plans for nonuniformed employees were quite sound. Since the eligibility requirements were tight, it was unlikely that the pension rolls would grow too large; and as the retirement allowances were modest, it was unlikely that the pension costs would come to very much. With the members contributing a sizable portion of their salaries to the plans—and the cities putting in enough money to cover long-term costs (and in a few cases to amortize prior obligations)—it was likely that the revenues would be adequate to pay future bills and prevent the buildup of large deficits. These plans worked out more or less as expected and, with a few exceptions,[53] have not put much of a strain on the big cities.

The firemen's and policemen's pension plans were another matter entirely. From a fiscal standpoint, their eligibility requirements, retirement benefits, contribution rates, and funding arrangements left much to be desired. How much was not clear at the start. But it became evident when several changes took place that devastated these systems and left in their wake a fiscal problem that still plagues most of the nation's big cities.

CHAPTER TWO

Swelling Rolls and Soaring Costs

From the outset, advocates of firemen's and policemen's pensions denied charges that the proposed systems would place a heavy burden on the taxpayers. A few even argued that the system would break even and perhaps run a small profit. Most of their fellow Americans were inclined to go along with them. Underlying this optimistic projection was the assumption that, although pension rolls would grow and pension costs would rise as the systems matured, they would never add up to very much. Not even the staunchest opponents of these pensions foresaw a time when the number of retired officers and survivors would approach, much less equal, the number of active officers or expected that one day pension payrolls would come to one-third or more of active payrolls.

This assumption seemed to hold up fairly well in the late nineteenth and early twentieth centuries. One reason is that the cities had few uniformed officers. New York, with by far the largest department, had only 3,300 policemen in 1890. In the other cities with more than 200,000 people, the size of the force ranged from over 1,600 to under 200. For every 1,000 residents, New York had

2.2 policemen. Most of the other cities had more than one, and two had fewer than one policeman for every 1,000 residents. The fire departments were even smaller. New York, again with the largest force, had only 1,000 fire fighters in 1890. In the other big cities, the size of the force ranged from more than 900 to fewer than 200. For every 1,000 residents, Buffalo had 1.2 firemen. But only two of the other cities with more than 200,000 people had more than one fireman for every 1,000 residents.[1]

Not only were there few firemen and policemen, but many left the service without a pension. Local politicians regarded the fire and police departments as prime sources of patronage. The officers had little in the way of job protection, and a change in municipal administration was often followed by a purge of the municipal payroll. When the Democrats returned to power in Chicago in 1897, Superintendent Joseph Kipley demoted and dismissed so many Republican policemen that, in the words of the *Chicago Tribune*, "There was not a Republican holding a place higher than desk sergeant, and so few of them they would not figure in a village census." The Democrats on the Los Angeles police force did not fare any better when the Republicans recaptured City Hall in 1889. In most cities the turnover was not as high in the fire departments. But in Kansas City, the home of the Pendergast machine, the department underwent several purges in the early twentieth century. Following shifts in political control, roughly one officer out of every four left the service in 1912, in 1916, and again in 1918, most of them involuntarily. Of the nearly 500 fire fighters who left the department between 1908 and 1929, fully 60 percent received no pension benefits at all.[2]

Another reason the assumption seemed to hold up is that the firemen and policemen who qualified for pensions did not remain on the rolls very long. To begin with, the average age at which officers joined was quite high. At the time, few cities imposed strict age requirements on candidates for the fire and police departments. The Los Angeles police force and the New Orleans fire department took applications from anyone under fifty. Forty was the maximum age requirement in the Philadelphia police force and

Baltimore fire department. Hence many recruits were in their thirties when they entered the service; some were in their forties and fifties. As a result, the average entry age was about thirty. Between 1866 and 1882, Cleveland's police recruits averaged just under thirty-two years of age. The average entry age was thirty in St. Louis in the 1880s and thirty-one in Baltimore in the 1900s. Fire fighters started at a slightly younger age. Recruits averaged twenty-eight in Boston in the 1870s and thirty in Baltimore in the 1880s.[3]

Moreover, the average age at which officers retired was rather high. To qualify for a pension, an officer had to work for twenty or twenty-five years, depending on the city, and serve until fifty or sixty. As a result, most firemen and policemen put in a good many years on the force and were well along in years when they retired. The Boston firemen and policemen who joined the service in the 1870s and were on the pension rolls in 1914 averaged thirty years of service and were, on average, fifty-nine at retirement. The service pensioners on New York City's rolls in 1914 averaged twenty-seven years on the force and retired, on average, at fifty-five. The situation was much the same in Baltimore and Philadelphia.[4]

Finally, the life expectancy of retired officers was not very high. At the turn of the century, the life expectancy of a fifty-five-year-old white male was seventeen years; at sixty it was fourteen years. For firemen and policemen it was slightly lower. According to the New York City Commission on Pensions, whose actuary based his analysis on the experience of retired officers between 1908 and 1914, a fifty-five-year-old officer who retired on a service pension would live fifteen years. If he quit at sixty, he had twelve years to go. The life expectancy of a disabled officer was slightly lower. A study done for the Buffalo City Council came up with similar results.[5] If the New York City and Buffalo mortality rates were typical, the retired firemen and policemen spent close to two years on the force for every year on the pension rolls.

With few firemen and policemen to begin with, with many leaving the force without a pension, and with the rest spending two years in service for each year on retirement, it is no wonder that the pension rolls grew slowly in the late nineteenth and early twentieth centuries. By 1914, several decades after its pension systems were set up, New York City had only one beneficiary for every three firemen and policemen. Chicago, whose pension systems were established in the mid-1870s, had one beneficiary for every four officers. The situation was much the same in Buffalo, Washington, D.C., and Boston. The pension rolls grew even more slowly in other cities. As late as the mid-1920s, there was only one beneficiary for every five firemen and policemen in San Francisco, one for every eight in Detroit, and one for every eleven in Los Angeles.[6]

With the retired firemen and policemen receiving half-pay at most and with their survivors receiving a good deal less, the pension costs grew fairly slowly, too. New York City, which had by far the largest firemen's and policemen's pension payroll, paid $3.5 million in benefits in 1914. Nowhere else but in Chicago did the pension payroll exceed $1 million in 1915. Only in New York did the pension payroll reach 15 percent of the active payroll; and only in Chicago and Cleveland did it come to more than 10 percent. A look at Chicago reveals just how modest an impact firemen's and policemen's pensions had on municipal expenditures. In 1915, the city spent nearly twice as much on public health as on pensions, nearly four times as much on refuse collection, and fully ten times as much on streets and highways.[7]

By the turn of the century, however, several changes were under way that would bring about a sharp rise in the pension rolls. The growth of the uniformed forces was one of these changes. As early as the late nineteenth century, many fire and police chiefs began to complain that their forces were severely understaffed and to press the authorities for additional officers. Citing the growing demands on the Washington, D.C., police, Superintendent Samuel H. Walker pleaded with Congress in 1886 for 100 more officers, a large increase for the capital's small force. Pointing out that "the already thin blue line is daily becoming thinner and thinner," Commissioner William F. Russell urged the Chicago City Council in 1928 to authorize 2,000 additional officers. These requests were usually supported by the Progressives (and later by the upper middle and upper classes in general) as a way to upgrade policing and fire fighting.[8]

The effort to expand the uniformed forces generated a good deal of opposition. Some reform groups argued that the police and fire departments left a lot to be desired not because they had too few officers but because they did not get the most out of them. Political interference and improper organization were the problems, not inadequate manpower. This position was shared by some property owners' groups, which held that the cities could not afford to take on additional policemen and firemen. Before placing so heavy a burden on the taxpayers, elected officials should insist that the forces make better use of available personnel. But the opposition labored under severe handicaps. Many residents were so troubled by the spread of crime and lawlessness, which reached a peak in the 1920s, that they were highly responsive to pleas for more policemen. Some property owners believed that additional fire fighters would improve the quality of fire protection and thereby bring down the costs of fire insurance. Many elected officials realized that they had to expand the uniformed forces just to maintain the existing level of service. Not only were the cities growing rapidly in the early twentieth century, but at the same time most fire and police departments were reducing manpower by shortening the workday and workweek, lengthening annual vacations, and extending sick leaves.[9]

And expand they did. Between 1890 and 1930, the police forces more than doubled in eighteen of the twenty largest cities; they tripled in some and grew even faster in others. The fire departments expanded at about the same pace. In many cities, however, the expansion of the uniformed forces was largely offset by the increase in population. Thus by 1930, only two of the twenty largest cities had more than 2.5 police officers for each 1,000 residents, and none of the cities had more than two firemen for each 1,000 residents. The expansion of the uniformed forces came to a halt in the 1930s, but picked up again after World War II, especially in the mid- and late 1960s. Between 1930 and 1970, a time when many big cities were growing very little or not at all, the police forces grew in all but one of the twenty largest cities, more than doubling in half of them. The fire departments also expanded, though less rapidly.[10]

By 1970, the uniformed forces were larger than ever. In the lead was New York City, with nearly 31,600 policemen. Washington, D.C., had more than five policemen for each 1,000 residents—more than twice as many as any city in 1890. Only one city had fewer than two officers per 1,000 people, which had been the case in all but one city in 1890. Much the same was true of fire fighters. New York City, with by far the largest force, had nearly 13,900 firemen. Boston had nearly three firemen for every 1,000 residents—more than twice as many as any city in 1890. And none of the twenty largest cities had fewer than one fireman per 1,000 people, which had been the case in all but a handful of cities in 1890.[11]

Even more important to the Progressives than enlarging the fire and police departments was removing them from partisan politics, which a New York City magistrate described in 1905 as "the curse of our free institutions." Politics, the reformers contended, had a baneful effect on the uniformed forces. Holding their jobs at the whim of the politicians, most officers placed a higher priority on pleasing the ward bosses than on fighting fires and chasing criminals; they served the machines rather than the public. Often forced out of office when the administration changed hands, the uniformed officers could not develop the skills necessary to cope with the rising levels of crime and conflagration in urban America. Applying the military analogy, the reformers insisted that the fire and police departments "should be as free of the unwholesome influence of partisan politics as the army and navy."[12] To get rid of political interference, they recommended that the fire and police departments be put under civil service. Each officer should be appointed on the basis of merit and hold office on good behavior, removable only for cause and after the presentation of formal charges. Not until the firemen and policemen got out from under the thumb of the local politicians would they be able to provide the level of service their fellow Americans expected.

The reformers' proposal ran into a lot of resistance. Ward leaders feared that they would no longer have a say over the appointment and removal of firemen and policemen. If they could not remove uncooperative officers and replace them with party loyalists, the politicians would lose much of their leverage over the uniformed forces and much of their clout in the ethnic communities. Many firemen and policemen objected to changing a system that provided them access to well-paying jobs. Also opposed were many lower- and lower middle-class immigrants, from whose ranks most of the uniformed officers were drawn. Here, too, the opposition was at a disadvantage. The Progressives had plenty of money, lots of power, and a good deal of prestige. The periodic scandals in the police and fire departments strengthened the reformers' case that political interference lowered the quality and raised the costs of public service. And many fire and police chiefs joined the reform effort on the ground that it would be impossible to upgrade policing and fire fighting unless the rank and file were protected from partisan politics.[13]

In the end, the Progressives won out. New York and Boston put firemen and policemen under civil service in the 1880s. Milwaukee and Chicago followed their lead a decade later. By about 1915, all but a handful of the big cities provided that firemen and policemen hold office on good behavior and be removed only for cause. The civil service procedures made it very hard to get rid of uniformed officers except for gross malfeasance—and not always then. By the 1920s, Raymond B. Fosdick and other reformers who had originally favored civil service came to regard it as "a bulwark for neglect and incompetence." Other bulwarks were erected after 1930. The Los Angeles voters approved a charter amendment that moved the discipline of policemen from the Civil Service Commission to a departmental trial board and imposed a one-year statute of limitations on alleged offenses. The Portland electorate adopted a charter amendment that made it harder to dismiss police officers than other municipal employees. Not long after, the rank-and-file associations prevailed on the authorities in New York and other cities to set up formal grievance procedures at which accused officers had the right to be represented by counsel and to cross-examine witnesses. By the late 1960s, the civil service rules and grievance procedures were so formidable that in the view of many Americans they unduly restricted the chiefs' authority to get rid of incompetent officers.[14]

The changes in disciplinary procedures brought about a marked decline in the number of dismissals. Only 7 percent of the Washington, D.C., firemen who left the service between 1943 and 1956 were fired. Only 8 percent of the Cincinnati policemen who quit the force between 1935 and 1960 did so involuntarily. The rate was even lower in Los Angeles. Although a good many officers resigned, the large majority of firemen and policemen eventually got a pension. More than three-fifths of the Detroit policemen who left the service between 1930 and 1950 went on the pension rolls. So did three of every four Oakland officers who were appointed after 1947. And more than nine of every ten New York City firemen who left the service between 1971 and 1979 went on the pension rolls.[15]

The reform movement brought about another change that had a profound, if unintended, impact on the pension rolls. From the outset, the reformers were deeply troubled by the advanced age of many of the recruits. In the absence of stringent age requirements, the reformers argued, the fire and police departments attracted many candidates who had failed in some other line and regarded public service as a last resort. Many of the recruits were already so fixed in their ways, so hostile to new ideas, and so full of misconceptions that they could not be trained to be competent policemen and fire fighters. Well into their thirties (and in some cases even into their forties and fifties), many of them would be too old to carry out their arduous and dangerous duties in ten or fifteen years. If they worked their way up the ranks—a slow process under civil service—they would probably be too worn out to assume command responsibilities. Insisting that the optimal age for a recruit was between twenty-one and twenty-five, the reformers brought strong pressure on the authorities to disqualify any applicant who was more than thirty. A few even recommended that the maximum age requirement be set as low as twenty-seven.[16]

The effort to tighten the age requirements encountered a good deal of opposition at the start. Speaking against a proposal to reduce the maximum age for police recruits from forty to thirty-five, Alderman James Power of Boston argued that many forty-year-olds are in "the prime of life." Alderman Hugh O'Brien claimed that age alone was not a proper basis for excluding anyone from the police department, pointing out that "some men are as young at fifty as others are at thirty-five." Following a spirited debate, the Boston City Council narrowly turned down the proposal in 1875. But in time, the campaign gathered momentum. During the late nineteenth and early twentieth centuries, most Americans came around to the view that policing and fire fighting were quasi-military activities that required exceptionally fit and vigorous men. They were troubled by reports that many departments employed officers in their sixties and seventies and that some kept "old and incapacitated men tucked away in every corner." Many Americans supported reformers like Smedley D. Butler, an ex-Marine Corps general who took over as Philadelphia's public safety director in 1924 and soon after urged the Philadelphia Civil Service Commission to reduce the maximum age for recruits from thirty-eight to thirty.[17]

Starting in the late nineteenth century, one city after another tightened the age requirements and in some cases imposed them for the first time. By the mid-1880s, Boston turned away applicants to the fire and police departments who were more than thirty years old—although later on it raised the age limit to thirty-five. Also in the vanguard were New York, which set the limit at thirty for policemen and twenty-nine for firemen, and Detroit, which fixed it at thirty for both services. Between 1915 and 1933, Portland reduced the maximum age for police recruits from fifty to thirty-six. Oakland, which accepted firemen up to thirty-five in 1917, cut them off at twenty-nine in the 1950s. By the 1960s, most cities imposed a maximum age requirement of thirty-five, and many turned away applicants who were more than thirty.[18]

The results were striking. During the late nineteenth and early twentieth centuries, the average entry age of Cleveland policemen dropped from thirty-two to twenty-seven. For Boston firemen it fell from twenty-eight to twenty-five. The trend persisted into the second third of the twentieth century. The average entry age of Baltimore policemen dropped to twenty-six after World II; in the Milwaukee fire department it fell to twenty-four. From the turn of the century to the late 1950s, the average entry age of Oakland firemen declined from thirty-one to twenty-four; for Oakland policemen it fell from thirty to twenty-four. Much as the reformers hoped, policing and fire fighting became, in the words of a high-ranking Washington, D.C., official "a young man's work." And a young man's work it has remained.[19]

Few reformers gave much thought to the impact their efforts to reduce the entry age would have on the pension funds. Those who did were confident that the change would work to their benefit. If the recruits joined the force in their late thirties or early forties, the reformers pointed out, it would not be long before they were eligible for retirement. In view of the hazards of policing and fire fighting, older officers were more likely to be disabled in the line of duty than younger ones. The longer an officer remained on the force, the more he would contribute to the pension fund and the more it would earn on his contributions.[20]

This analysis made sense—but only if the plans had stiff eligibility requirements. If a city imposed an age requirement of sixty for service pensions, as Los Angeles did in 1899, it was obviously in the interests of the pension system for the recruits to join as young as possible. An officer who joined the Los Angeles police force at twenty-five would serve ten years longer than an officer who entered at thirty-five and would contribute at least 40 percent more to the pension fund. The same held true for disability pensions, provided that the cities gave them only to officers disabled in the line of duty, carefully screened applications, and put the burden of proof on the officers.

During the late nineteenth and early twentieth centuries, however, the uniformed forces brought strong pressure on the authorities to ease the eligibility requirements—in particular to abolish the age requirements for service pensions. In the face of claims that most officers were pretty much worn out after twenty or twenty-five years on the job, the authorities usually gave in. By 1909, firemen could retire at any age after serving twenty years in New York and twenty-five years in Cleveland. Policemen were eligible for a pension regardless of age after putting in twenty years in Pittsburgh and twenty-five in Detroit. The Louisiana legislature abolished the age requirement for New Orleans policemen in 1912. And despite objections that uniformed officers would be able to retire in the prime of life, the Los Angeles City Council eliminated the age requirement in 1919. Thus by 1929, a great many cities, including New York, Detroit, Pittsburgh, and Cleveland, allowed firemen and policemen to retire at any age after twenty to twenty-five years of service. Several others—among them Chicago, Minneapolis, and Philadelphia—set the age requirement for service pensions at forty-five or fifty.[21]

The firemen and policemen also prevailed on the authorities to loosen the eligibility requirements for disability pensions. Following the lead of New York City, Philadelphia and Cleveland gave pensions not only to officers who were disabled in the line of duty but also to officers who were disabled in any manner after serving for ten years. So many big cities followed suit that by the late 1920s the great majority of uniformed officers who had put in five or ten years were eligible for a non-service-connected disability pension. After World War II, Massachusetts, Pennsylvania, New York, and a few other states passed what were known as heart and lung laws. Under these laws, public officials were required to presume that a fireman or policeman who came down with heart or lung trouble incurred it in the line of duty and was entitled to a service-connected disability pension. The requirements were also loosened in Washington, D.C., when Congress passed a law in 1962 that provided that an officer was eligible for a duty-disability pension if an injury incurred on or off duty was later aggravated on duty. Under the so-called aggravation clause, a district fire chief was granted a disability pension in 1978 on the basis of a back injury suffered in a fall in 1943.[22]

The changes in eligibility requirements had a profound impact.

Instead of serving twenty-five to thirty years before retirement, firemen and policemen put in, on average, twenty to twenty-five years. As they joined the force in their mid- or late twenties, they retired in their early or mid-fifties (or even late forties). The Los Angeles firemen and policemen who went out on service pensions in the 1930s and 1940s put in, on average, twenty-two years and retired at forty-nine. And the New Orleans firemen on the pension rolls in 1954—the great majority of whom were service pensioners—put in, on average, twenty-four years and retired at fifty-one. The average years of service and average age at retirement were much the same in New York and Detroit. And they were even lower in Washington, D.C., where firemen and policemen who retired between 1966 and 1971—more than 90 percent on a disability pension—averaged twenty-two years of service and retired at forty-six.[23]

Not only did firemen and policemen retire earlier, but as the years passed they also lived longer. At the turn of the century an officer who went on the pension rolls at fifty would live about eighteen more years. By the early 1930s, however, a study of the Los Angeles Fire and Police Pension System found that an officer who retired at fifty would live roughly another twenty-two years. George B. Buck came up with similar findings in a study of Cincinnati firemen and policemen. So did the New York State Commission on Pensions, which analyzed mortality rates of New York City firemen and policemen in the mid-1930s, and the U.S. Treasury Department, which looked into mortality rates of Washington, D.C., firemen in the mid- and late 1950s. A subsequent Treasury Department study, based on the experience of District firemen and policemen from 1966 to 1971, showed that an officer who retired at fifty would live nearly twenty-three more years. A more recent San Francisco study indicated that the life expectancy of firemen and policemen has gone up since then.[24]

The rise in life expectancy of firemen and policemen was partly the result of the overall increase in longevity in the twentieth century. Although particularly pronounced among women and blacks, the increase extended to white men, who have always made up the great majority of firemen and policemen in urban America. From the turn of the century to the early 1970s, the life expectancy of a white male at fifty went up from 20.8 to 23.6 years. Also contributing to increased life expectancy was the improved health of fire and police recruits. During the late nineteenth century, most cities imposed minimal physical and health requirements on candidates. As the medical examiners owed their jobs to the local politicians, a well-connected candidate had little trouble getting around the requirements. Many recruits were alcoholic and overweight; some suffered from more serious ailments. But early in the twentieth century, the Progressives prevailed on the authorities to tighten the requirements and to insulate the examiners from political pressure. Henceforth the recruits were at least as healthy as other young men, perhaps even healthier.[25]

With life expectancy rising and average entry age and years of service falling, the firemen and policemen spent (or would spend) about as many years on the pension rolls as on the job. On average, the Los Angeles officers who retired on a service pension in the late 1930s and 1940s served for twenty-two years and received benefits for twenty-three. The New Orleans firemen on the pension rolls in 1954 served for twenty-four years and stayed on the rolls for twenty-two. The situation was much the same in Detroit, where the service pensioners in 1965 would spend more than four years on the rolls for every five years on the force. It was even worse in Washington, D.C., where the officers who left the service between 1966 and 1971 would spend four years on the pension rolls for every three years on the job.[26]

Moreover, the pensions did not necessarily stop when the retired officers died. Consider the experience of the Oakland firemen and policemen who went out on service pensions between 1951 and 1975. More than nine of every ten officers, virtually all of whom were male, were married at retirement. As women live longer than men, fully seven of every ten retirees were survived by their wives. Unless they remarried, which was rare, the widows were entitled to a minimum of two-thirds of their late husbands' benefits. Although the retired officers, on average, died at seventy-five, the widows went on the pension rolls at an average age of sixty-four.

They came on the rolls so early not only because they were, on average, five or six years younger than their husbands, but also because they were much more likely to survive the retired officers who died in their fifties and sixties than the ones who died in their seventies and eighties. At sixty-four, the widows had an average life expectancy of eighteen years. Thus, after putting in twenty-six or twenty-seven years, the typical Oakland fireman or policeman collected a pension for about nineteen years; his widow received another pension, perhaps with somewhat lower benefits, for another eighteen years. For every three years the officer put in on the job, he and his widow spent four years on the pension rolls.[27]

With a great many firemen and policemen to begin with, with most of them leaving the service with a pension, and with many officers spending as much time on the rolls as on the job, the pension rolls grew far beyond expectations. As figure 2.1 shows, the growth was striking in New York, Washington, D.C., Detroit, and Los Angeles. The growth was rapid in other cities as well. Be-

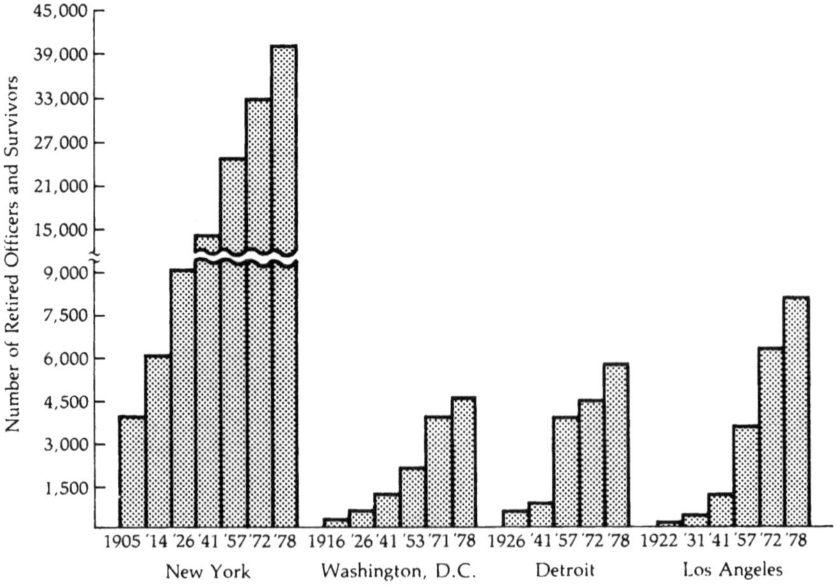

FIGURE 2.1
Growth of Pension Rolls in Selected Cities

tween 1915 and 1941, the pension rolls more than doubled in Buffalo and more than tripled in Chicago. Between 1941 and 1972, they went up more than two times in Denver, nearly three times in Indianapolis, and almost four times in Atlanta. During the 1970s, the firemen's and policemen's pension rolls grew at a rapid pace just about everywhere.[28] And as figure 2.1 reveals, they reached unheard of levels by the end of the decade. The pension rolls will continue to rise in the years ahead. The rise will probably be especially sharp in Washington, D.C., and several other cities in the late 1980s and early 1990s, when the many uniformed officers who were appointed in the aftermath of the 1960s riots become eligible for retirement.

The pension rolls not only went up; they went up more than the active rolls. As figure 2.2 reveals, the ratio of beneficiaries to officers climbed sharply in New York, Washington, D.C., Detroit, and Los Angeles. The ratio soared in other cities as well. Between 1915 and 1941, the pension rolls went up nearly four times as much

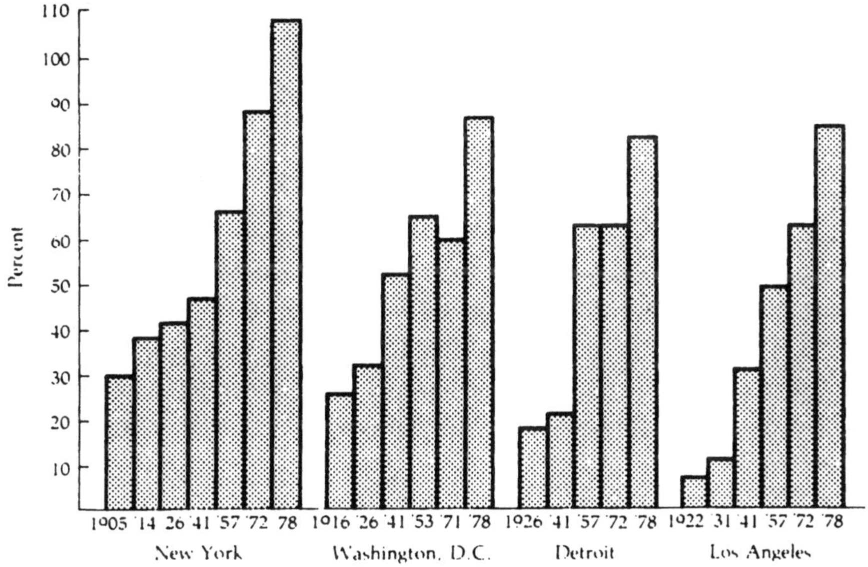

FIGURE 2.2
Retired Officers and Survivors as a Percentage of Active Officers in Selected Cities

as the active rolls in Buffalo and almost six times as much in Chicago. Between 1941 and 1972, the pension rolls rose nearly twice as much as the active rolls in Pittsburgh, more than four times as much in Portland, and more than six times as much in St. Louis.[29]

The trend continued during the 1970s. Thus by the end of the decade, the ratio of beneficiaries to officers was higher than ever. It was also much higher in the firemen's and policemen's pension systems than in the retirement plans for nonuniformed employees, where the turnover rates were so high that many employees left their jobs without a pension and where the eligibility requirements were so stiff that most employees spent far more time on the job than on the pension rolls. At a time when few cities had as many as one beneficiary for every two nonuniformed employees, St. Louis had nearly two beneficiaries for every three uniformed officers. There were more than three for every four in Portland, more than four for every five in Detroit and Los Angeles, and more than six for every seven in Washington, D.C., and New Orleans. Oakland and San Francisco had as many persons on the pension rolls as on the active rolls. New York City had the unenviable distinction of having more retired officers and survivors than active officers.[30] Outside the Southwest, most big cities had two fire departments and two police departments, one made up of active officers and another, somewhat smaller but growing more rapidly, made up of retired officers and survivors.

Not only did the pension rolls swell, but the retirement allowances soared. One reason was that the cities liberalized the pension benefits. Through the early twentieth century, retired officers received modest allowances. Nowhere did policemen get more than half their salary. Firemen did so in Boston and Cleveland—but only under special circumstances. A few cities even gave the retired officers less than half-pay. Some cities gave roughly half-pay to an officer's widow and children, but most paid sharply reduced benefits to the survivors unless the officer died in the line of duty. And some gave them no benefits at all.[31]

The situation began to change after World War I. At the urging of the Milwaukee Pension Laws Commission and the Milwaukee police force, the Wisconsin legislature pegged retirement benefits to years of service and raised the maximum allowance for retired officers from 50 to 75 percent of final salary. Two years later, the legislature passed a similar law for Milwaukee firemen. Following an agreement worked out by the Los Angeles Fire and Police Protective League and the city's commercial and civic groups, the voters adopted a charter amendment in 1922 that hiked maximum benefits from one-half to two-thirds of final salary. The amendment also provided that on the death of a retired officer his widow would receive one-half his final salary. Other cities followed suit after World War II. At the behest of the Oakland firemen and policemen, the voters approved a charter amendment in 1941 giving the widows of retired officers a full pension if their husbands died of injuries incurred in the line of duty. If they died in other ways, the widows would get two-thirds of their pensions. Six years later, the Washington, D.C., police and fire departments, whose spokesmen claimed that they were having trouble attracting and retaining qualified officers, prevailed on Congress to raise the maximum benefits for disability pensions from 50 to 70 percent of final salary.[32]

By the mid-1970s, an officer who retired after thirty years of service received 80 percent of final salary in Washington, D.C., more than 70 percent in several cities, and more than 60 percent in most others. In several cases the benefits were higher if he served longer. An officer who was disabled in the line of duty got over 50 percent of final salary in all but a handful of cities, 60 to 80 percent in most of them, and up to 90 percent in Los Angeles, Milwaukee, and San Francisco. For non-service-connected disability, the benefits were generally far lower. If an officer died in the line of duty, his widow collected at least 50 percent of his final salary in all but a couple of cities, 60 to 80 percent in many others, and 100 percent (plus his contributions) in Baltimore. Some cities provided additional benefits for children under eighteen. When an officer died of ordinary causes, the benefits were, as a rule, much lower. If an officer died after he retired, his widow received all or most of his pension in Los Angeles, New York, and Chicago; roughly half as much in several other cities; and, in the rest, the same as she would have if the officer had died before retirement.[33]

Another reason the retirement allowances soared was that the cities raised the salaries on which the allowances were based. Through the early twentieth century, officers' salaries were fairly low. In 1915, maximum salaries for patrolmen ranged from $1,400 a year in San Francisco to $900 in New Orleans. Maximum salaries for fire fighters were about the same. What is more, fire and police salaries increased very little over the years. In 1915, patrolmen earned $1,400 in New York, up only $200 since 1866, and $1,000 in Baltimore, up only $100 since 1878. In New Orleans, policemen actually earned less in 1915 than in 1868. Much the same was true of fire fighters.[34] In other words, through the early twentieth century, most uniformed officers began and ended their careers at about the same or only slightly higher salaries.

But once again things started to change after World War I. Caught up in an inflationary spiral that had just about doubled the cost of living since 1914, firemen and policemen put enormous pressure on the authorities to raise salaries. Their spokesmen claimed that, with wages lagging far behind prices, most officers were forced not only to cut back on necessities but also to go into debt; even then many of them could not make ends meet. More often than not, these demands were supported by reform groups, which believed that higher salaries were essential in order to attract qualified recruits. Yielding to this pressure, the authorities approved sizable salary hikes in one city after another. Between 1915 and 1929, maximum salaries for policemen doubled in Detroit and Los Angeles, went up more than 90 percent in Cleveland and Minneapolis, and climbed more than 80 percent in Philadelphia and Chicago. Maximum salaries for fire fighters increased at about the same rate. On the eve of the Great Depression, the rank and file earned over $2,000 in most cities and over $2,500 in a few.[35]

With the onset of the Depression, most cities held the line on wages and in some cases lowered them. But after World War II, firemen and policemen, claiming that wages were falling so far behind prices that they could not support their families, began again to press for higher salaries. They pushed so hard that, between 1940 and 1960, maximum salaries for the rank and file nearly tripled in Baltimore, Milwaukee, Los Angeles, and San Francisco and more than doubled in most other big cities. The uniformed officers kept up the pressure in the 1960s and 1970s. Exploiting widespread fear of crime, arson, and public disorder, and capitalizing on the near-record rate of inflation, they made impressive gains. Between 1960 and 1978, maximum salaries more than tripled in Philadelphia, Detroit, and Washington, D.C., and more than doubled elsewhere. By the end of the decade, veteran patrolmen and fire fighters earned between $15,000 and $20,000 in most big cities and even more in a few. The wage hikes were so large that a San Francisco patrolman who retired in 1972 after twenty-five years of service would have received a pension worth more than three times his starting salary and more than his base salary in the first twenty years.[36]

Yet another reason the retirement allowances soared was that many cities gave the retired officers cost-of-living adjustments. Through the early twentieth century, virtually all retirees received a fixed pension—that is, their allowances did not fluctuate with changes in salaries, benefits, or the cost of living. A Boston ladderman who retired in 1900 on half-pay of $600 a year would still have gotten $600 in 1913, even though salaries by then had gone up to $1,300 and a ladderman who retired that year on half-pay would have gotten $650 a year. From time to time, the authorities granted ad hoc postretirement increases. Faced with a situation in which most Boston firemen and policemen who retired before 1892 received one-third pay and most who retired after 1892 got half-pay, the Massachusetts legislature gave all retired officers half-pay in 1893.[37] But this action was exceptional. Most retired officers collected the same allowance year after year, an arrangement that was satisfactory as long as the cost of living rose very slowly or not at all.

But this, too, began to change after World War I. In 1923, after several years of severe inflation, Detroit policemen prevailed upon the voters to adopt a charter amendment that pegged retired officers' allowances to police salaries. Every time salaries were raised, allowances would go up half as much. Two years later, the voters approved the same arrangement for firemen. Congress was just as generous. At the behest of the District of Columbia's uniformed forces, it passed the Equalization Act of 1923, which provided that retired officers' pensions go up at the same rate as active officers' salaries. The uniformed officers succeeded in getting a postretirement escalator in Denver, but, despite vigorous campaigns, they were thwarted in Chicago and New York City. Detroit, Washington, D.C., and Denver remained the exceptions until well after World War II, when firemen and policemen elsewhere, alarmed by the declining purchasing power of their pensions, started to push for a hedge against inflation. The Minnesota legislature gave the Minneapolis firemen and policemen a cost-of-living escalator in 1953. At the prodding of the Fire and Police Protective League, Los Angeles approved a postretirement adjustment of 2 percent a year for service pensioners in 1966; three years later, the voters extended it to disability pensioners; and in 1971, they removed the 2 percent ceiling, pegging benefits to the consumer price index (CPI). The San Francisco firemen and policemen won an escalator a few years later.[38]

By the mid-1970s, firemen and policemen had some sort of postretirement escalator in slightly over half of the twenty-nine largest cities. Los Angeles and Seattle pegged pensions to the CPI. Several other cities linked benefits to the CPI but imposed a ceiling. For example, St. Louis limited the adjustment to 3 percent a year and to 25 percent in a lifetime. Denver, Indianapolis, San Francisco, and Washington, D.C., pegged pensions to salaries. The escalators drove the allowances way up. A Los Angeles fire fighter who retired in 1967 on a service pension of $5,700 a year received $11,200 in 1979—8 percent more than his final salary; a Washington, D.C., patrolman who retired in 1965 on a disability pension of $6,000 a year received $14,300 in 1977—68 percent more than final salary.[39]

All of these changes had a profound impact. Spurred by the sharp hike in fire and police salaries after the outbreak of World War I, retirement allowances rose a good deal in the 1920s. But the rise came to a halt with the onset of the Great Depression, which forced many cities to hold the line on wages or even to reduce them. Retirement allowances increased much faster after World War II, especially in the 1960s and 1970s, when most cities raised fire and police salaries at a record clip and many set up postretirement escalators. From 1941 to 1972, the average benefits for retired officers climbed from about $1,600 to nearly $6,600 in Los Angeles and from under $1,300 to more than $6,600 in Detroit. They went up almost as much in most other cities.[40]

Fueled by large wage hikes and double-digit inflation, retirement allowances rose even faster in the 1970s. Thus by the end of the decade, not only were firemen's and policemen's retirement allowances higher than ever, but they were also much higher than the allowances of nonuniformed employees, whose salaries were much lower, whose benefits were less liberal, and whose payments were not pegged to wage hikes or to the cost of living. At a time when few retired nonuniformed employees received more than $5,000 a year, and many less than $4,000, retired officers collected, on average, over $6,000 in Chicago, over $7,000 in St. Paul, over $8,000 in New York, and over $9,000 in Portland. Other cities were even more generous. Retired officers received, on average, more than $10,000 a year in Los Angeles, more than $12,000 in San Francisco, and close to $15,000—roughly 80 percent of a patrolman's or fire fighter's base pay—in Washington, D.C.[41]

Survivors' allowances went up too. Between 1941 and 1972, they climbed from nearly $500 to close to $2,600 a year in Minneapolis and from just under $1,400 to more than $7,100 in Los Angeles. At the same time, average stipends went up three times in Chicago and more than five times in Washington, D.C. Survivors' allowances have gone up even faster since 1972. Their allowances varied widely from city to city, however, much more so than the officers' allowances. At one extreme were St. Louis and Pittsburgh, where the widows received less than $2,000 a year in 1977, roughly one-third as much as the retired officers. At the other extreme were Los Angeles and San Francisco, where widows got more than $10,000 and $12,000, respectively, about as much as the retired officers. As the survivors made up a fair share of the pension rolls—ranging from 7 percent in Oakland to 50 percent in Chicago, and from 20 to 40 percent elsewhere—the rise in their retirement allowances contributed a good deal to the growth of the pension payrolls.[42]

As pension rolls swelled and retirement allowances climbed, pension payrolls soared. As figure 2.3 shows, the increase was enormous in New York, Washington, D.C., Detroit, and Los Angeles.

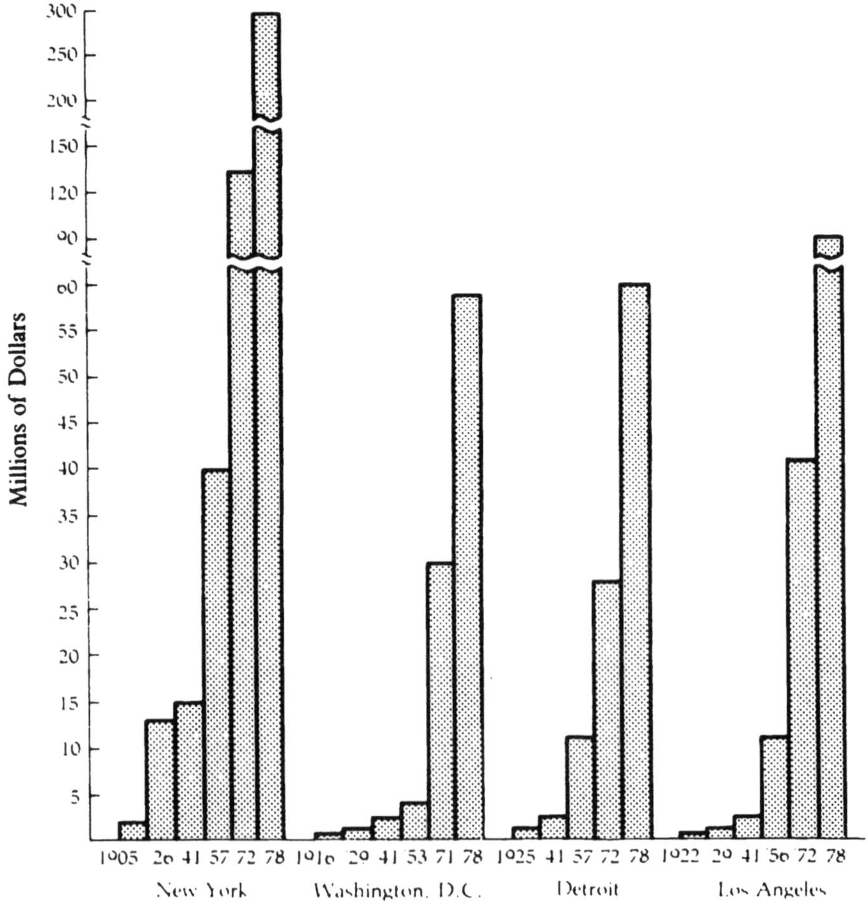

FIGURE 2.3
Firemen's and Policemen's Pension Payrolls in Selected Cities

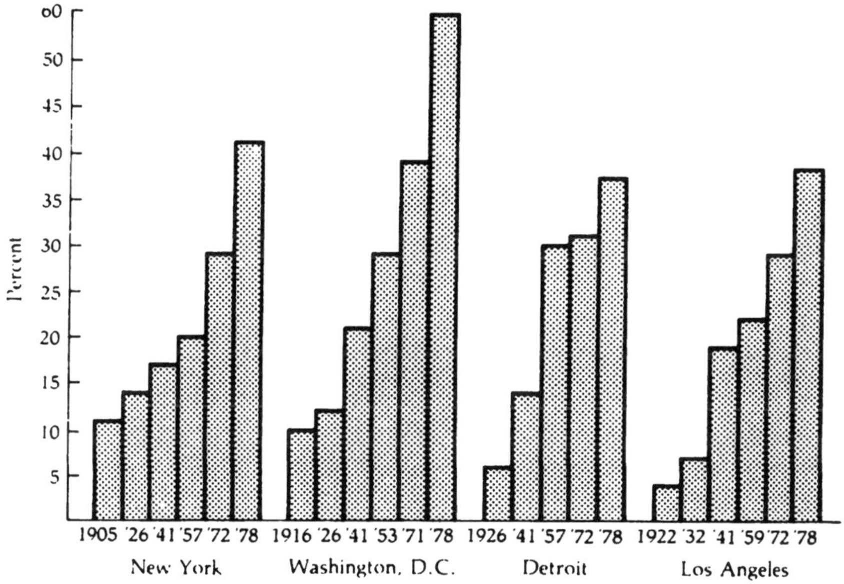

FIGURE 2.4
Firemen's and Policemen's Pension Payrolls as a Percentage of Active Payrolls in Selected Cities

It was impressive in other cities as well. Between 1915 and 1941, the pension payrolls went up nearly sixfold in Minneapolis and more than ninefold in New Orleans. Between 1941 and 1972, they rose nearly five times in Pittsburgh, more than eight times in Atlanta, and more than thirteen times in Denver. Pension payrolls continued to climb during the 1970s.[43] And as figure 2.3 reveals, they reached astonishing levels by the end of the decade. If pension rolls continue to swell and retirement allowances continue to climb—both of which are highly likely—pension payrolls will probably go up at least as much in the 1980s as they did in the 1970s.

Pension payrolls not only went up; they went up more than active payrolls. As figure 2.4 shows, the ratio of the pension payroll to the active payroll soared in New York, Washington, D.C., Detroit, and Los Angeles. It climbed in other cities as well. Between 1915 and 1941, the pension payroll rose more than twice as much as the active payroll in Chicago and more than four times as much in Buffalo. Between 1941 and 1972, it went up close to three times as much in Indianapolis and more than three times as much in New Orleans.[44]

Pension payrolls continued to go up faster than active payrolls in the 1970s. Thus by the end of the decade, the pension payroll came to more than 20 percent of the active payroll in Denver and more than 30 percent in New Orleans. It exceeded 35 percent in Detroit and Los Angeles, 40 percent in New York and Oakland, and approached 60 percent in San Francisco and Washington, D.C.[45] By contrast, pension payrolls came to less than 15 percent of active payrolls in most retirement plans for nonuniformed employees, which had far fewer beneficiaries per employee and paid them far smaller allowances. In most cities, firemen's and policemen's pension payrolls will go up as fast as active payrolls in the years ahead. How these cities, already hard pressed to support one fire department and one police department, will manage to support two fire departments and two police departments remains to be seen.

CHAPTER THREE

Lagging Revenues and Mounting Deficits

At the outset, most Americans were confident that the cities would have no trouble paying the firemen's and policemen's pension bills, which they were assured would not amount to very much. The authorities shared this confidence. Rather than give the pension systems a chunk of the property taxes, which would have brought in a good deal of money, they gave them only the proceeds from several special sources. Some of the revenues came from the day-to-day operations of the uniformed services. Chief among them were fines levied on policemen and firemen for breaking department regulations, rewards given officers by grateful citizens, receipts from sales of unclaimed merchandise, and fines imposed on residents who violated municipal ordinances. Other revenues came from assessments on business interests that benefited directly from the activities of the uniformed services or that required more than an ordinary degree of their attention—principally taxes on fire insurance premiums and licenses for saloons, pawnshops, and secondhand stores. To supplement these sources, firemen and policemen paid dues; held balls, fairs, and other fund-raising affairs; and in some cases contributed a small portion of their wages.[1] Through the nineteenth century, most Americans felt sure that these revenues would cover the pension costs.

Their confidence, it turned out, was misplaced. As early as 1913, the Bureau of Municipal Research reported that the New York City police pension fund was in serious trouble. With the swelling of the rolls after 1900, revenues had risen much slower than expenses, which reached $2.1 million—roughly 15 percent of payroll—in 1912. The resulting deficits, which came to more than $1.1 million that year, severely depleted the fund. To meet its obligations, the city had started to make direct appropriations to the system in 1903; a decade later, these appropriations came to almost 42 percent of total revenue. Despite this financial transfusion, the pension fund was, in the bureau's words, "a legal fiction only," with virtually no assets on hand. Three years later, the New York City Commission on Pensions released a study that not only confirmed the bureau's findings about the police pension fund but revealed that the firemen's relief fund was in serious trouble, too. For years, revenues had lagged behind expenses, which climbed from roughly $370,000 in 1900 to more than $1 million—about 14 percent of payroll—in 1914. Plagued with recurrent deficits, the fund was gradually being depleted, dropping from more than $1.3 million in 1900 to just over $850,000—only $200,000 less than its annual expenses—fourteen years later.[2] The fund would have been on the verge of insolvency if the city had not begun making direct appropriations in 1912.

The situation was just as bad in other cities. Testifying about Washington, D.C.'s firemen's and policemen's pension system, Deputy-Auditor Daniel J. Donovan told the House District Committee that revenues had fallen so far behind expenses that the authorities had been forced to prorate benefits in 1910. Retired firemen received 69 percent of their allowances, retired policemen 59 percent. Several years later, actuary J. D. Craig informed the Buffalo City Council that since 1910 the policemen's and firemen's pension funds had been paying out more than they had been taking in. The deficits were gradually depleting the funds, which had on hand only a small fraction of the assets needed to meet future obligations. After a thorough study, the Illinois Pension Laws Commission reported in 1917 that the state's pension systems were virtually insolvent. In worst shape of all were Chicago's firemen's and policemen's pension funds, which would have to come up with an additional $106 million in the years ahead to pay the retirement benefits of current members and their dependents. The same year, the New Jersey Pension and Retirement Funds Commission found that the firemen's and policemen's pension systems were running deficits in Newark and several other cities—"a most appalling situation," the commission declared. Looking at this evidence, some observers warned that America's cities might soon be in the same bind as London and Paris, where pensions were currently consuming 20 to 30 percent of payroll.[3]

Summarized in pamphlets put out by the Bureau of Municipal Research and reports issued by the Carnegie Foundation, these findings aroused intense concern throughout the nation, particularly among the many upper middle- and upper-class organizations that spoke for urban America's business and civic elites. Although these groups often favored public pensions, they were committed to keeping a lid on property taxes. They feared that the firemen's and policemen's pension systems, if left unchecked, would generate a steep, and altogether unpredictable, rise in municipal expenditures. Typical of these groups was the Municipal League of Los Angeles. Upon learning of the New York City Pension Commission's findings, the league insisted in 1917 that the City Council hire an actuary to do a study of firemen's and policemen's pension costs, which at the time came to only 3 percent of payroll. Pointing out that pensions abroad had reached 20 and even 30 percent of payroll, the league served notice on the council two years later that it would oppose changes in eligibility requirements or retirement benefits until the pension system was put on a sound financial basis.[4]

According to George Buck, Lewis Meriam, and Paul Studensky, the leading authorities on public employee pensions in the early twentieth century, a sound financial basis meant two things. It meant that public employee pension systems should be contributory—that employees and employers should each bear a fair share, preferably one-half, of the costs. A corollary of the principle that pension systems served both groups, a contributory plan had several practical advantages, Studensky insisted. It would generate a good deal of revenue and thereby reduce the taxpayers' burden, which might otherwise become prohibitive. It would also check extravagant demands by the employees, who would have to pay part of the costs. A sound financial basis also meant that public pension systems should be funded—that enough money should be put aside to pay for benefits as they accrued and not when they came due. Every year, the authorities should appropriate a sum which, when added to the employees' contributions and prudently invested, would be enough to meet the system's long-term obligations. A funded system would prevent one generation from passing pension bills on to another generation, Meriam argued. It would also reveal the full costs of the pension systems, reduce the fluctuations in annual appropriations, and, in the long run, enhance the solvency of the pension funds.[5]

By these criteria, the firemen's and policemen's pension systems were unsound. In most cities, they were wholly noncontributory. And in some cities, the uniformed officers made what might be called token contributions. In New York City, policemen put 2 percent of salary into the pension fund, but from 1893 to 1914 their contributions came to less than 16 percent of total revenues. And in Chicago, where policemen had 1 percent taken out of their salaries, employee contributions made up roughly 14 percent of total revenues between 1910 and 1915. Employee contributions fell far short of one-half of current costs in all but a few systems. In 1915, the New Orleans firemen's system took in only one-third as much from active officers as it paid out to retired officers; the Detroit policemen's fund took in less than one-sixth as much. In Cleveland, where uniformed officers were assessed from 50 cents to $1.25 a month, depending on rank, employee contributions covered only

4 percent of the costs of the policemen's pension system and just 3 percent of the costs of the firemen's pension fund.[6]

The pension systems were unfunded as well. They operated on a pay-as-you-go basis—paying the pension benefits when they came due rather than as they were accrued—drawing on fines and fees and, if need be, tapping general revenues. Some systems built up small funds in their early years, when very few officers were on the pension rolls, but as the rolls swelled and the costs soared, the funds were depleted and the systems built up sizable unfunded liabilities. The problem was particularly acute in New York and Chicago. According to the New York City Commission on Pensions, the policemen's fund had a deficit of $74 million in 1914—nearly $7,000 an officer—and the firemen's fund had a deficit of $41 million—more than $8,000 for each officer. The Illinois Pension Laws Commission found that the Chicago policemen's plan had unfunded liabilities of $30 million in 1916, roughly $6,000 per officer, and the firemen's plan had unfunded liabilities of $13 million, nearly $7,000 for each officer. If these systems remained unfunded, it was estimated that the authorities would in time have to put in 35 to 45 percent of payroll in New York and 34 to 37 percent in Chicago to pay the annual pension bills.[7]

Out of the concern aroused by these findings emerged a host of efforts to put the firemen's and policemen's pension systems on a sound financial basis. Commercial and civic groups took the lead. Buck, Meriam, Studensky, and A. A. Weinberg (probably the foremost public sector actuary in the second third of the twentieth century) helped out. So did the New York Bureau of Municipal Research, the Detroit Bureau of Governmental Research, and other reform organizations. After World War II, several statewide pension or retirement commissions lent their support as well.[8] The principal objectives of these efforts were to make the pension systems contributory, with firemen and policemen bearing close to half the costs, and to place the systems on a funded basis, with the authorities required to meet the long-term costs and to amortize the unfunded liabilities. Unless these changes were put into effect, their advocates believed, the systems' revenues would fall behind their expenses, their assets would gradually be depleted, and their deficits would climb to intolerable levels.

The firemen and policemen got deeply involved in the efforts to put their pension systems on a contributory basis. It was, after all, their take-home pay and their pension funds that were at stake. Representing the policemen and firemen were a host of benevolent and fraternal associations, most of which had been founded in the late nineteenth and early twentieth centuries. The associations spoke for the uniformed officers not only on pensions but also on salaries, disciplinary procedures, and working conditions. Although many of the firemen's organizations were affiliated with the International Association of Fire Fighters, chartered by the American Federation of Labor in 1918, they were not labor unions. Not until the 1950s and 1960s, long after the first attempt to unionize the police had collapsed in the wake of the Boston police strike of 1919, were these groups recognized as bargaining agents for the rank and file.[9] But although the benevolent and fraternal associations were not labor unions, they were well-organized pressure groups. And by dint of skillful lobbying and sizable campaign contributions, they built up a good deal of clout with the elected officials who set the contribution rates for the big-city firemen's and policemen's pension systems.

In some cities, firemen and policemen strongly objected to putting more money into the pension funds or, if the funds were noncontributory, to making contributions at all. New York City is a case in point. Under pressure from civic groups, Mayor Jimmy Walker appointed a committee to study the firemen's and policemen's pension systems in 1927, ten years after the Commission on Pensions first called attention to the pension problem. Over the opposition of representatives of the firemen and policemen, the committee recommended that henceforth the uniformed officers put 4 percent of salary into the pension funds.[10] Antion Holterback, president of the Uniformed Firemen's Association, whose members had a noncontributory system, denounced the proposal, branding it "an insult" to the force. Stressing the extreme hazards of fire fighting, he insisted that firemen not be required to pay for pensions. Pointing out that the rank and file were woefully underpaid, he argued that a 4 percent reduction in salary would be a terrible hardship. Joseph Moran, president of the Patrolmen's Benevolent Association, whose members put in 2 percent of salary, voiced similar objections. Joining forces, the two associations headed off the proposal for a couple of years. And when it was incorporated into a bill in 1931, they managed, with the help of upstate firemen and policemen, to bottle it up in committee.[11]

In other cities, firemen and policemen were willing to contribute to their pension funds. Los Angeles is a good example. Early in 1925, the Municipal League of Los Angeles launched another in a long series of attacks on the city's firemen's and policemen's pension system, which was then noncontributory. Following a meeting with the Los Angeles Fire and Police Protective League, which had been organized by the uniformed officers in 1922, it set up a committee that examined the pension system and recommended that it be made contributory. The Protective League was far from enthusiastic about pension reform. But fearful that an intransigent stance might jeopardize its efforts to win a wage hike for the uniformed forces, it proposed that firemen and policemen contribute 2 percent of salary for service pensions—a proposal later endorsed by the rank and file. Not long after, the league told the City Council that the officers would pay half the costs of service pensions, even if it came to more than 2 percent of salary, but that the public should bear the full costs of disability pensions. Pleased with the Protective League's position, commercial and civic groups backed the wage hike, which was approved by the voters in April 1926. The league then went along with a proposed amendment to the city charter that required the firemen and policemen to contribute 4 percent of salary and otherwise tightened the pension system. The amendment was adopted in November 1926.[12]

Milwaukee is another good example. Late in 1920, the Milwaukee Pension Laws Commission released a long-awaited report urging the authorities to set up a pension system for nonuniformed employees, a proposal that went nowhere, and to overhaul the firemen's and policemen's pension plans, which were in precarious condition. At the heart of the report were recommendations to tighten the eligibility requirements, raise the retirement benefits, put the funds on a reserve basis, and hike the contribution rate from 2.5 to 4.6 percent—to which the city would add an additional 13.75 percent. Some firemen and policemen opposed the recommendations on the grounds that the contribution rate was too high and favored asking the state legislature not to revise the pension systems but to give them more money instead. Aware that the funds would soon be depleted, and doubtful that the legislature would bail them out, other officers favored the commission's recommendations. After much debate, the City Council shelved the commission's plan for the firemen but—over the objections of the Citizens' Bureau, which considered a system into which the employer put three times as much as the employees "much too liberal"—adopted the commission's plan for the policemen. At the urging of the Milwaukee police, the state legislature passed an enabling act in July 1921. The City Council accepted it four months later. Shortly after, the firemen changed their minds. At their request, the authorities established a new plan that closely resembled the policemen's and included a provision that hiked the contribution rate from 2.5 to 4.75 percent.[13]

Similar changes took place in other cities. Between 1915 and 1931, the policemen's pension systems went on a contributory basis in Boston and Pittsburgh, as did the firemen's systems in Baltimore and San Francisco. The contribution rate for policemen rose from 1 to 3 percent in New Orleans and from 2 to 4 percent in Buffalo. For firemen it climbed from 1.5 to 3.5 percent in Washington, D.C., and from 1 to 4 percent in Portland. With the conspicuous exceptions of New York and Detroit firemen, most uniformed officers were making contributions by the late 1930s; a good number were putting in more than ever. Nonetheless, many of the pension systems were only nominally contributory. Policemen put in only 2 percent in Baltimore, Minneapolis, and Seattle. Firemen also contributed 2 percent or less in many cities. Only in Chicago, Milwaukee, Oakland, and, in the case of firemen, St. Louis did uniformed officers contribute more than 4 percent. Nowhere did they put in more than 5 percent.[14]

By the late 1930s, and in some cities even earlier, the firemen and policemen were forced to reconsider their position vis-à-vis the contribution rate. At a time when many nonuniformed employees were putting 5 percent or more into their pension funds, which were much less liberal, it was hard to justify nominal contributions, much less noncontributory pensions. As pension costs soared, moreover, the firemen and policemen were hard pressed to resist demands from civic and commercial groups for tighter eligibility requirements and lower retirement benefits. Their spokesmen also realized that as long as the pension funds were noncontributory or nominally contributory it would be hard to prevail upon the authorities to liberalize the systems.

Out of this reconsideration emerged a more sophisticated position—one that was closely adhered to after World War II. Instead of attempting to preserve the remaining noncontributory systems or to keep the contribution rates as low as possible, the firemen and policemen offered to put more money into the funds in order to head off what they saw as drastic reforms and to strengthen their proposals for off-duty disability pensions, improved survivors' benefits, and cost-of-living escalators.

The uniformed officers took this tack in Detroit in 1940. Long troubled by the high costs of firemen's and policemen's pensions, the Bureau of Governmental Research and other reform groups called on the authorities to include the uniformed forces in the city's general employees' retirement plan, which had tighter eligibility requirements, lower retirement benefits, and higher contribution rates. Stressing the extreme hazards of their work, the firemen and policemen rushed to the defense of their pension plans. In an effort to resolve the conflict, the City Council asked the reformers and the officers to work out a compromise. They soon reached an impasse. When the council went on record as opposed to consolidation, the uniformed officers drafted a new plan. The plan retained the existing eligibility requirements and added an off-duty disability pension but, as a sweetener, raised the contribution rate to 5 percent—up from 1 percent for policemen and 2 percent for firemen. Although the reformers came up with a plan of their own, which treated uniformed officers much the same as nonuniformed employees, the council incorporated the firemen's and policemen's proposal into a charter amendment and put it on the November ballot. Speaking in favor of the proposal, former Police Commissioner James K. Watkins stressed that the 5 percent contribution would generate an additional $500,000 a year and over time save the taxpayers $7.5 million. The firemen and policemen campaigned hard for the new system. Despite objections by the Bureau of Governmental Research, which insisted that the amendment was unduly liberal, the voters approved it by a majority of nearly two to one.[15]

Events unfolded in a similar way in Los Angeles. By the mid-1940s, many firemen and policemen felt that their pension system left much to be desired. An officer received no benefits for non-service-connected disability, they complained; and as a result of a charter amendment approved in 1927, he had to put in twenty-five years to be eligible for a service pension. This dissatisfaction was skillfully exploited by the Police Employees Union—a labor organization established in 1943 with the help of the American Federation of Labor—which soon emerged as a threat to the Fire and Police Protective League. With the support of Mayor Fletcher Bowron and Chief C. B. Horrall, however, the Police Commission broke the union in 1946. To solidify its position, the Protective League then submitted to the City Council a charter amendment that created an off-duty disability pension, lowered the service requirement from twenty-five to twenty years, and, to pay for the changes, raised the contribution rate from 4 to 6 percent. The council put the proposal on the May 1947 ballot. The Chamber of Commerce, which had replaced the Municipal League as the watchdog of the city's pension funds, opposed the amendment. So did the Board of Pension Commissioners. But stressing that the hike in contributions would save the taxpayers roughly $2 million over the next twelve years, the Protective League persuaded the voters to endorse the amendment by a majority of close to three to two.[16]

A similar strategy was employed in Washington, D.C. Following a major revision in 1956 of the Civil Service Retirement System, which covered most nonuniformed federal employees, the District of Columbia firemen and policemen mounted a campaign to improve their pension plan. Working with representatives of the District commissioners and the fire and police chiefs, they drafted a bill that incorporated most of their demands. It loosened eligibility requirements, raised retirement benefits, and provided, for the first time, a non-service-connected disability pension for officers who had put in at least five years. To cover the costs, the bill hiked the contribution rate, which had been raised from 3.5 percent to 5 percent in 1949, to 6.5 percent. Alvin E. Davis, president of the Firemen's Association, and Police Chief Robert F. Murray defended the bill on the grounds that it would help the uniformed forces attract and retain qualified officers, a position endorsed by several citizens groups. Alfred A. Capasso, president of the Policemen's Association, added that as a result of the increased contributions the bill would entail "no immediate cost" to the government. Despite strenuous objections by Senator Harrison Williams, who argued that the proposed changes would raise pension costs to nearly 40 percent of payroll, with the government putting in more than five times as much as the officers, Congress adopted most of the provisions of the original bill in 1957.[17]

Similar changes took place in other cities. Following the lead of Detroit, New York City put its firemen's pension fund on a contributory basis in 1940; and later on Columbus, Ohio, and Memphis, Tennessee, set up contributory systems for uniformed personnel. The contribution rates also went up. Between the mid-1930s and the mid-1970s, for example, the rate for policemen rose from 2 to 6 percent in Seattle and from 4.75 to 9 percent in Chicago; for firemen it climbed from 3 to 6 percent in Pittsburgh and from 2 to 7 percent in St. Louis. By the mid-1970s, all the big-city firemen's and policemen's pension systems were contributory—

although by virtue of collective bargaining agreements reached in 1970, Milwaukee paid all of the firemen's contributions and all but 1 percent of the policemen's. The rates ranged from 3 percent in Indianapolis to 9 percent in Chicago, with most officers elsewhere putting in 5 to 7 percent.[18]

Spurred by the hike in contribution rates, the rise in wages, and the growth of the uniformed forces, employee contributions soared after World War II. As figure 3.1 shows, the increase was quite sharp in New York, Washington, D.C., Detroit, and Los Angeles. Employee contributions went up at a rapid rate in other cities as well. Between 1941 and 1977, they shot up tenfold in St. Louis, twentyfold in Portland, and even more in Chicago and Atlanta. They have continued to go up since.[19]

In a few cities, employee contributions went up faster than pension payrolls. But in most cities, as figure 3.2 reveals, contributions barely kept up with payments.[20] Thus by the late 1970s—more than half a century after Buck, Meriam, and Studensky first made the case for contributory pensions—firemen and policemen paid one-half of the pension bills only in a few cities. Their contributions came to less than 30 percent in Minneapolis, less than 20 percent in New Orleans, less than 15 percent in San Francisco, and less than 10 percent in Indianapolis. Firemen and policemen paid a much lower proportion of the pension bills than nonuniformed employees, whose contributions came to more than 50 percent of the pension payroll in most cities. In most cities, they also put much less money into the pension funds than the authorities. For every dollar the officers put into the funds, the authorities poured in more than $5 in New York, more than $7 in Washington, D.C., more than $12 in Detroit, and nearly $13 in Indianapolis. The employee contributions came to much more in the retirement plans for nonuniformed personnel, into which most cities put less than $3 for every dollar contributed by the employees.[21]

Despite the gradual rise in contribution rates, firemen and policemen have thus far paid only a small fraction of the pension bills in most big cities. Los Angeles is a case in point. Between 1928, when firemen and policemen began to contribute to their pension

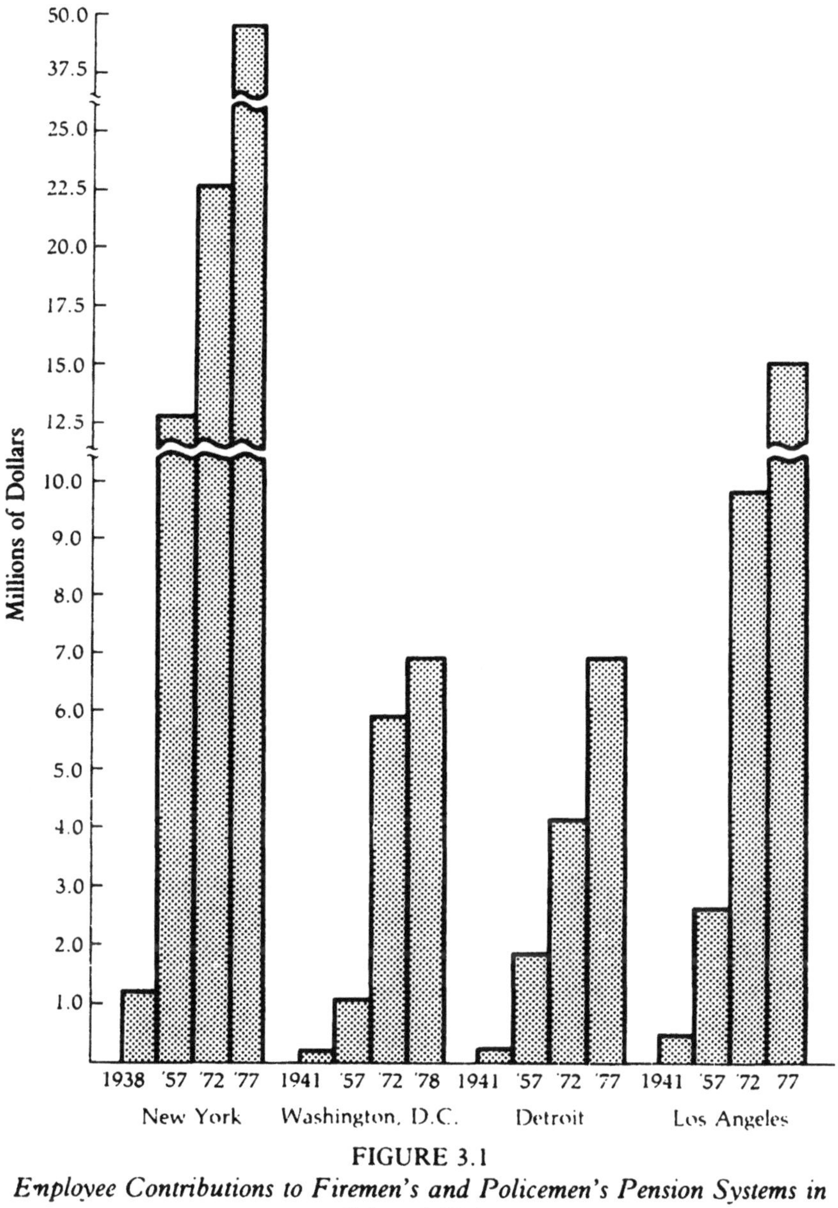

FIGURE 3.1

Employee Contributions to Firemen's and Policemen's Pension Systems in Selected Cities

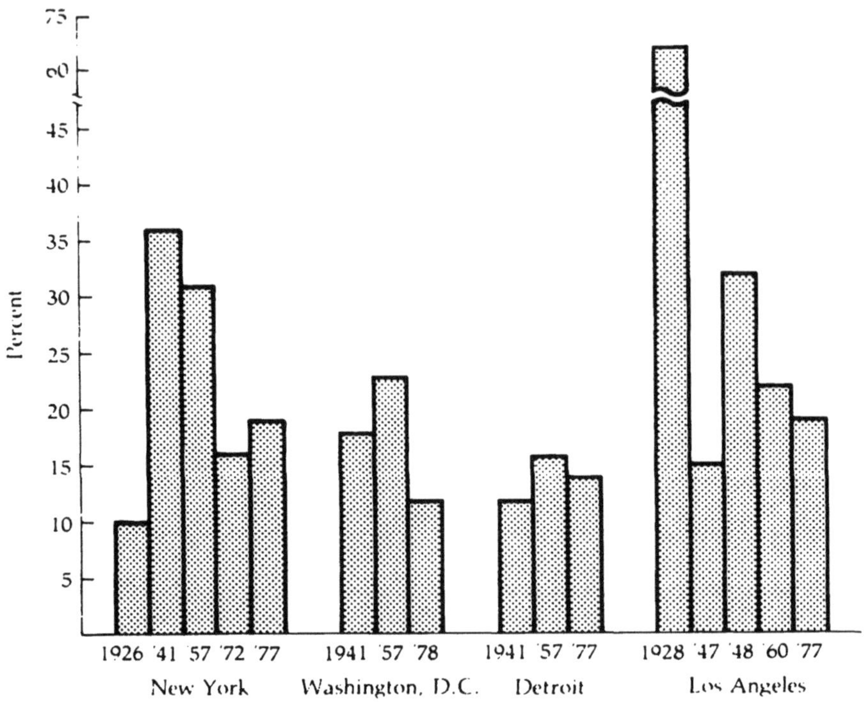

FIGURE 3.2
Employee Contributions as a Percentage of Firemen's and Policemen's Pension Payrolls in Selected Cities

system, and 1978, by which time the rate had gone up from 4 to 7 percent, employee contributions came to only 19 percent of pension payrolls. The situation was much the same in Detroit, where in 1941 the voters hiked the contribution rates from 1 to 5 percent for policemen and from 2 to 5 percent for firemen. Even so, employee contributions made up only 20 percent of pension payments from 1941 to 1978. In much the same bind was Oakland, where in 1951 the voters raised the contribution rate from 5 to 7 percent. Yet between 1951 and 1978, the firemen and policemen paid for only 19 percent of the pension benefits.[22]

Firemen and policemen will continue to pay only a small fraction of their pension bills in the future, too, especially in cities that have operated on a pay-as-you-go basis. According to the Washington, D.C., Office of Budget and Management Systems, a fireman who retired in 1973 after twenty-one and a half years of service would have put $11,000 into the pension system. Assuming normal life expectancy and a 5 percent annual hike in salaries, to which his retirement allowance was pegged, he and his survivors will eventually receive $547,000. In other words, his contributions will cover less than 2 percent of his benefits; for every dollar he put into the system, the authorities will have to pay out more than $50. According to the University of Washington's Bureau of Governmental Research, things were much the same in Seattle. A policeman who retired in 1966 after twenty-five years of service would have paid $4,500 into the pension system. Again assuming normal life expectancy and a 3 percent annual rise in salaries, to which his retirement allowance was tied, he will in time collect $118,000. His widow will get an additional $24,000. His contributions will cover only 3.4 percent of his benefits.[23] Although the percentages would be higher in cities that run on a reserve basis and earn interest on employee contributions, it is nonetheless fair to say that many big-city firemen's and policemen's pension systems are still only nominally contributory.

Employee contributions lagged behind pension payrolls for several reasons. One was that, at the insistence of the uniformed officers, the authorities usually set the rates too low to cover anything like half the system's costs. In 1921, the Wisconsin legislature passed an enabling act that put the Milwaukee police pension fund on a reserve basis and raised the contribution rate from 2.5 to 4.75 percent. The Citizens' Bureau protested that at that rate the officers would pay only one-quarter of the pension bills. But the City Council, afraid that the policemen would oppose the changes if the contribution rate were set any higher, accepted the act. Shortly after, it approved a similar plan for firemen. Twenty years later, Detroit's firemen and policemen drafted a charter amendment that put their pension system on a funded basis and hiked the contribution rate from 1 or 2 to 5 percent. At that rate, member contributions would have covered only a small fraction of pension bills. But the City Council, realizing that pension reform was doomed without the support of the uniformed forces, placed the amendment on the ballot, where it was approved by a large majority.

Another reason contributions lagged behind expenses was that, under pressure from the uniformed forces, the authorities often changed the pension plans in ways that increased pension costs without raising contribution rates. In 1961, New York State eliminated a provision that New York City policemen had to work for ten years to become eligible for a non-service-connected disability pension. Eight years later, the legislature raised the benefits for firemen who were disabled in the line of duty. Despite the changes, the contribution rates remained the same. In 1968, San Francisco approved a charter amendment that hiked the maximum allowance for retired officers from 70 to 75 percent of final salary and redefined final salary from average salary in the last three years to average salary, overtime included, in the last year. Although the changes were estimated to cost $3 million a year, the contribution rate stayed the same. In 1970, Congress gave in to pressure from the Washington, D.C., firemen and policemen to allow them to retire after twenty years, regardless of age. It also raised the retirement allowances. Although Congress hiked the contribution rate from 6.5 to 7 percent, the extra revenue was not nearly enough to offset the additional costs.[24]

Probably the most costly of these changes was the provision of postretirement escalators. The Minneapolis escalator, which was mandated by the state legislature in 1953, pegged retirement allowances to current salaries of veteran patrolmen and fire fighters. Their salaries went up so much between 1967 and 1975 that average payments jumped from $3,200 to $8,500 a year for policemen and from $3,800 to $8,400 a year for firemen. If an officer retired in 1967, his retirement allowance would probably have exceeded his peak earnings by 1975—and just about doubled his average earnings. The Los Angeles escalator, adopted in 1966 and liberalized five years later, tied retirement benefits to the consumer price index (CPI). For this, the employees contributed 1 percent of salary. The CPI rose so rapidly that by 1978 the city had to shell out 26 percent of fire and police payroll to pay for the escalator—slightly more than it paid for all the system's other benefits. The San Francisco escalator, approved in 1974, provided that current benefits be based on 1975 salaries and that future benefits be increased by one-half of wage hikes. For this, the employees made no contribution. The year after the escalator went into effect the pension payroll soared from $20 million to $31 million and the employee contributions dropped from 21 to 13 percent of pension costs.[25]

Yet another reason contributions lagged behind costs was that the authorities gave the firemen and policemen hefty raises after World War II—especially in the 1960s and 1970s. To understand why, it is instructive to look at the case of a Los Angeles patrolman who joined the force in 1952 and retired twenty-five years later. His starting salary was only $4,300 a year, but as a result of a series of wage hikes his final salary came to nearly $19,000—roughly twice as much as his average salary. This difference is critical, because the officer's contributions, taken out of his paycheck every month for twenty-five years, were based on average salary, whereas his benefits were based on final salary. The disparity between final salary and average salary would not have been as wide in the case of a policeman or a fireman whose wages went up at a slower pace and whose benefits were based on average salary in the last three years of service. But it would have been even wider in the case of a superior officer whose contributions were based largely on a patrolman's or fire fighter's salary but whose benefits were based on the salary of a sergeant, lieutenant, or captain, which was a good deal higher.[26]

Troubled by the way wage hikes drove up pension bills, a few cities worked out an ingenious agreement with the policemen and firemen. Provided that the officers lowered their wage demands, the cities would pick up part of their pension contributions. The arrangement appealed to the cities because it held down the officers' final salaries, on which their retirement allowances were based. The officers favored the scheme because it raised after-tax earnings. On the basis of such an agreement, Milwaukee picked up all of the firemen's 7 percent contribution in 1969 and all but 1 percent of the policemen's 7 percent contribution a year later, thus giving the officers virtually noncontributory pensions. Under the Increased-Take-Home-Pay program, which got under way in the early 1960s, New York City paid 2.5 and later 5 percent of the policemen's and firemen's contributions, which were 7.2 and 5.3 percent, respectively. The program's results were disappointing. Not only did wages go up steadily, but between 1963 and 1972, a time when the fire department payroll climbed from $89 million to $170 million, the firemen's contributions dropped from $5.4 million to less than $500,000. The policemen's contributions went up, but only slightly.[27] Although the state legislature cut back the program in 1976, it still cost the city much more than it saved the taxpayers.

The firemen and policemen had mixed feelings about the attempts to put their pension systems on a funded basis. They realized that a pay-as-you-go plan could not be counted on to meet future obligations. Yet if the pension systems went on a funded basis, the actuaries would have to start computing long-term costs and the cities would have to begin paying them. Once one generation of taxpayers learned that the costs could not be passed on to the next generation, they would probably insist that the firemen and policemen put more money into the pension plans. They might even take a closer look at the officers' frequent efforts to liberalize the pension systems.

The commercial, civic, and property owners groups were also ambivalent about the attempts to put the firemen's and policemen's pension systems on a funded basis. Deeply disturbed by the mounting liabilities, they feared that if the systems stayed on a pay-as-you-go basis it would not be long before the costs were prohibitive. Yet if the systems went on a funded basis, the cities would be compelled to make much larger appropriations, at least in the short run. As these groups were well aware, the taxpayers would have to bear the brunt of this burden. In other words, they would have to pay not only their own pension bills but also the pension bills left by previous generations of taxpayers.

The attempt to put the pension systems on a funded basis went fairly smoothly in Chicago. A few years after the Illinois Pension Laws Commission issued its 1917 report, the state legislature took under consideration a bill to overhaul the Chicago policemen's pension system. In addition to tightening eligibility requirements, raising retirement allowances, and providing non-service-connected disability pensions, the bill would have put the system on a reserve basis. To cover the costs, the policemen would have had to contribute 4.75 percent of salary to the fund (up from 2.5 percent) and the city would have had to appropriate an additional 10.5 percent of payroll. Chicago's civic groups supported the bill, as did the *Chicago Tribune*. The Chicago policemen endorsed the bill, too. They were well aware that their pension system had been in financial trouble for several years and that their leaders had been forced time and again to appeal to the legislature for additional money. To them the bill was a way to ensure the system's long-term solvency. As Captain (later Superintendent) Morgan Collins put it, the bill "places the fund on a basis where it will automatically take care of all requirements for pensions for all time to come."[28] With the civic groups and uniformed officers in agreement, the legislature passed the bill in 1921, making the Chicago policemen's pension plan one of the first funded systems for uniformed officers.

The campaign for funding ran into trouble in Los Angeles. Late in 1922, the City Council submitted to the voters a charter amendment that would have put the firemen's and policemen's pension plan on a reserve basis and would have tightened the eligibility requirements. The Municipal League, a strong advocate of funding, opposed the amendment on the grounds that the eligibility requirements were still too loose. But at the prodding of the Fire and Police Protective League and the Chamber of Commerce, the voters adopted it by a wide margin. Eight years later, an actuary recommended that in view of the system's mounting deficits the city should increase its contributions by 44 percent a year. Coming in the depths of the Depression, the recommendation so upset the Fire and Police Protective League and the Chamber of Commerce that they agreed to underwrite jointly another actuarial report. This report concluded that the city should raise its contributions by 100 percent. Well aware that the hard-pressed taxpayers would not tolerate such a hike, the Protective League, Chamber of Commerce, and real estate interests prevailed on the council to put on the ballot a charter amendment taking the system off the reserve basis. Despite the vigorous opposition of the Municipal League, which favored retaining the funded system, the voters adopted the amendment in 1932, thereby putting the pension system back on a pay-as-you-go basis.[29]

The campaign for funding ran into even more trouble in New York City. Early in 1933, the Citizens Budget Commission, which had been set up by the business community to keep tabs on the city's financial condition, issued a report showing that firemen's and policemen's pension payrolls had been rising at a rapid pace and currently cost the taxpayers more than $7 million a year. To reduce this burden, the commission recommended that the two systems be closed to new members, that current officers be required to contribute 5 percent of salary, and that new appointees be compelled to join the City Employees' Retirement System, a funded system that covered most of New York's nonuniformed work force. A bill incorporating these recommendations was subsequently introduced into the state legislature. The city's firemen and policemen—who had previously beaten back several attempts to raise their contribution rates, put their funds on a reserve basis, and fold their systems into the City Employees' Retirement System—saw the bill as yet another attempt to deprive them of their hard-won benefits. With the backing of fellow officers from upstate New York, they lobbied hard against the bill. Despite reports by the State Commission on Pensions revealing that the city's firemen's and policemen's pension systems had accumulated a whopping $395 million in unfunded liabilities, the bill failed to make it through the legislature.[30]

Chicago policemen were not the only uniformed officers whose pension plans were put on a funded basis in the 1920s and 1930s. Milwaukee established a reserve system for policemen in 1921 and for firemen a couple of years later. Cincinnati required all officers appointed after August 1, 1931, to join the city's new municipal employees' retirement system, a reserve system drawn up by George B. Buck under the auspices of the Cincinnati Bureau of Municipal Research. And on the basis of a charter amendment approved in 1932, San Francisco brought its firemen and policemen into the city and county employees' retirement system, which had operated on a funded basis since its inception in 1921. These cities were the exceptions, however. Cleveland's firemen and policemen withstood pressures from the Citizens' League in the 1920s to put their pension plans on a reserve basis. Detroit's uniformed forces also fought off efforts by the Bureau of Governmental Research in the 1930s to fold their pension plans into the city's general retirement system. Thus by the late 1930s, most big cities still operated on a pay-as-you-go basis.[31]

As the years passed, however, the opposition to funding weakened. The firemen's and policemen's pension systems were building up massive unfunded liabilities, some of which exceeded $100 million. Civic groups viewed the deficits as a confirmation of their fears that a pay-as-you-go policy would lead the cities to financial catastrophe. Firemen and policemen saw them as a sign that a nonreserve system might jeopardize the long-term solvency of the pension funds. Commercial and real estate interests regarded the deficits as evidence that a pay-as-you-go policy would eventually impose an intolerable burden on the taxpayers. Moreover, virtually all the retirement systems for nonuniformed employees were funded. The uniformed officers and elected officials were hard pressed to explain why firemen's and policemen's pension systems operated on a different financial principle. It was one thing to argue that firemen and policemen were entitled to less strict eligibility requirements and more liberal retirement allowances because of the extreme hazards of their work; it was another to point to something distinctive about the uniformed services that justified keeping their pension systems on a pay-as-you-go basis.

As the opposition weakened, the campaign for funding picked up momentum. Early in 1954, the actuaries informed Mayor Norris Poulson that the Los Angeles Fire and Police Pension System had run up a deficit of $169 million, a sevenfold increase since the voters took it off a reserve basis in 1932. Alarmed by the deficit, which he blamed on the pay-as-you-go approach, Poulson appointed a citizens committee to look into the pension problem. The committee, which included representatives of the Fire and Police Protective League and the Chamber of Commerce, agreed that the system should be put back on a reserve basis but disagreed about how it should otherwise be changed. Whereas most of the members favored imposing an age requirement and reducing survivors' benefits, the Protective League's representatives argued that the changes would make it harder for the uniformed forces to attract qualified personnel. Poulson called on the City Council to put the majority's recommendations on the ballot, but, under strong pressure from the Protective League, the council refused to do so. At the mayor's urging the committee regrouped and, by laying aside the issues of eligibility requirements and benefit levels, reached agreement that the city should put the pension system back on a reserve basis and amortize the unfunded liability over fifty years. These recommendations were incorporated into a charter amendment that the council placed on the ballot in May 1959 and the voters adopted by a majority of nearly three to one.[32]

Early in 1967, the Minnesota Interim Commission on Public Retirement Systems, a legislative commission set up in 1955, reported that the local firemen's and policemen's pension systems were "financially sick." Operating on a pay-as-you-go basis, they had accumulated unfunded liabilities of $116 million, of which $53 million was accounted for by the Minneapolis plans and $32 million by the St. Paul systems. Declaring that the problem "will not go away," the commission called on the legislature to close the systems to new members, put them on a reserve basis, and amortize the unfunded liabilities over forty years. To ease the burdens on the cities, it urged that the uniformed officers and the state government bear a larger share of the costs. The firemen and policemen objected. Citing the success of the Social Security system, they defended the pay-as-you-go approach; they also stressed that the alternative meant tying up tax dollars that were sorely needed for other purposes. The League of Minnesota Municipalities, which was so troubled by the mounting pension costs that it favored replacing the local systems with a statewide plan, endorsed the commission's proposals. After much debate, the legislature worked out a compromise known as the Guidelines Act of 1969. The act put the systems on a reserve basis but did not close them; and instead of requiring the cities to amortize the liabilities over forty years, it ordered them just to pay the interest on the deficit.[33]

Early in 1973, the Treasury Department reported that the District of Columbia's firemen's and policemen's pension system had built up a deficit of $540 million—roughly $85,000 for each officer. If the system remained on a pay-as-you-go basis, it would not be long before the District spent more on pensions than on salaries. Coming in the wake of other studies showing that 90 percent of the officers were retiring on disability pensions, the report helped launch a major campaign to overhaul the system. Led by Senator Thomas Eagleton and Representative Romano Mazzoli, the campaign drew strong support from District officials. Even the uniformed officers favored putting the system on a reserve basis, although they opposed tightening the eligibility requirements. For a while the campaign bogged down on the issue of who should pay the unfunded liabilities, which by 1976 exceeded $1.2 billion, or about $200,000 for each officer. But in 1978, after several rounds of hearings, Congress passed a bill that, in addition to tightening the eligibility requirements, put the system on a funded basis and ordered the federal government to pick up the liabilities for the officers who retired before home rule came to the District. President Jimmy Carter vetoed the bill on the grounds that it imposed too heavy a burden on the federal government. But a year later, when Congress reduced the burden, he signed a bill that required the District to pay the long-term costs and to cap, although not to amortize, a portion of the unfunded liabilities over twenty-five years.[34]

The firemen's and policemen's pension systems went from a pay-as-you-go to a reserve basis in many other cities. Under Mayor Fiorello H. LaGuardia, New York City closed its two pay-as-you-go plans in 1940 and set up two funded systems, known as Article IA and Article II, for new firemen and policemen. On the basis of a charter amendment adopted in 1941, Detroit replaced its firemen's and policemen's plans, both of which operated on a pay-as-you-go basis, with a single reserve system. Cleveland's firemen and policemen were transferred into a funded system in 1965, when the Ohio legislature abolished the local plans and established one statewide fund for public safety officers. So many big cities followed this route after World War II that by 1976 most firemen's and policemen's pension systems operated on a funded basis.[35] Since then, a few cities have joined the fold. With civic groups, actuarial firms, and state retirement commissions all pushing for funding, it may not be long before the holdouts abandon the pay-as-you-go approach.

The cities poured lots of money into the pay-as-you-go systems. Between 1941 and 1978, the government's contributions soared from $1 million to $52 million in Washington, D.C., which came to 52 percent of fire and police payroll. For every dollar the authorities spent on fire and police salaries in the late 1970s, they put twenty-five cents into fire and police pensions in Portland and fifteen cents into them in Denver. But these contributions only paid for current benefits. They provided no reserves for future obligations, which, according to the actuaries, will eventually reach 50 percent of payroll in Portland, approach 100 percent in Denver, and exceed 100 percent in Washington, D.C. With tremendous obligations and negligible assets, the pay-as-you-go systems have built up enormous deficits. The unfunded liabilities climbed in Washington from $45 million in 1941 to more than $1.2 billion in 1976—from $20,000 to $200,000 for each officer and from eight to thirteen times fire and police payroll. As table 3.1 shows, the situation was pretty bad in Portland, Denver, and Pittsburgh too.[36]

The cities poured far more money into the reserve systems. Detroit, which put $250,000 into its funded system in 1942, its first full year of operation, contributed $88 million in 1978, roughly 54 percent of payroll and more than twice as much as it spent on sanitation. Los Angeles, which put $6 million into its pension system

TABLE 3.1
Unfunded Liabilities of Pay-as-You-Go Firemen's and Policemen's Pension Systems in Selected Cities

City	Year	Total Unfunded Liabilities	Unfunded Liabilities per Officer	Ratio of Unfunded Liabilities to Payroll
Pittsburgh	1976	$130 million	$ 80,000	6:1
Denver	1978	$299 million	$132,000	7:1
Portland	1976	$233 million	$170,000	9:1
Washington, D.C.	1976	$1.2 billion	$200,000	13:1

in 1959, the last year on a pay-as-you-go basis, contributed $115 million in 1978, about 48 percent of payroll and slightly more than it spent on highways. Elsewhere the cities' contributions ranged from $11 million in Minneapolis to a whopping $332 million in New York City. In addition to paying for current benefits, these contributions helped to build up the funds' assets. Between 1942 and 1978, the Detroit plan's assets climbed from $1 million to more than $532 million—from one-twentieth to more than one-half the system's deficit. Between 1959 and 1978, the Los Angeles system's assets soared from $20 million to more than $623 million—from about two to seven times the pension payroll. By the late 1970s, other pension funds had assets ranging from $28 million in Minneapolis to close to $2.2 billion in New York City.[37]

As the funds' assets grew, their earnings rose. In Detroit, they climbed from virtually nothing in 1942 to more than $24 million in 1978—more than three times as much as the firemen and policemen contributed to the system. In Los Angeles, the earnings soared from under $700,000 in 1959 to nearly $39 million in 1978—more than one-third as much as the taxpayers put into the system. By the late 1970s, the firemen's and policemen's pension funds earned from a low of almost $1 million in Minneapolis to a high of more than $126 million in New York City. Combined with the steady growth in employee contributions and government appropriations, the sharp rise in investment earnings lifted revenues to record levels. In 1978, they reached more than $127 million in Detroit, more than twice the pension payroll, and more than $170 million in Los Angeles, close to twice the pension payroll. By the late 1970s, these revenues ranged from $13 million in Minneapolis to more than $420 million in New York City.[38]

In some systems, current revenues kept pace with future obligations and held down unfunded liabilities. Consider the Chicago firemen's and policemen's pension funds. Operating on a pay-as-you-go basis, the funds accumulated a deficit of $45 million by 1916, about $6,600 for each officer and more than four and a half times the active payroll. The policemen's fund went on a reserve basis in 1921, the firemen's fund a decade later. Although the unfunded liabilities have gone up since then—reaching $891 million, or close to $50,000 per officer, in 1978—they now make up only two and a half times payroll. Or take the Philadelphia firemen's and policemen's pension plans, which went on a reserve basis in the late 1950s, when they were consolidated with the city's general employees retirement system. As of 1976, the plans had unfunded liabilities of only $401 million, which was just $35,000 for each officer and only two and a half times payroll.[39]

But in several systems current revenues failed to keep pace with future obligations and thus to hold down liabilities. Between 1932 and 1959, when the Los Angeles Fire and Police Pension System was on a pay-as-you-go basis, the unfunded liabilities soared from $24 million to $310 million (about $44,000 per officer and six times the payroll). Yet between 1960 and 1978, when the system was on a reserve basis, the deficit jumped to nearly $2.1 billion (roughly $212,000 per officer and nearly nine times payroll). The Detroit firemen's and policemen's pension system, which was set up on a funded basis in 1941, started out with a deficit of roughly $22 million—under $5,000 for each officer and less than twice the active payroll. Yet by 1978, the deficit had reached $895 million—about $122,000 per officer and more than five times payroll. As table 3.2 reveals, things were not quite as bad in New York City, where the firemen's and policemen's pension plans went on a reserve basis in 1941. But they were even worse in San Francisco, which aban-

TABLE 3.2
Unfunded Liabilities of Funded Firemen's and Policemen's Pension Systems in Selected Cities

City	Year	Total Unfunded Liabilities	Unfunded Liabilities per Officer	Ratio of Unfunded Liabilities to Payroll
New York	1978	$3.1 billion	$ 84,000	4:1
San Francisco	1978	$307 million	$ 92,000	5:1
Oakland	1977	$233 million	$190,000	9:1

doned the pay-as-you-go approach in 1932, and Oakland, which opted for funding in 1951. The situation was much worse in the firemen's and policemen's pension systems than in the retirement plans for nonuniformed employees, most of which managed to hold the unfunded liabilities down to under $25,000 for each employee and less than twice the payroll.[40]

There are a few reasons many firemen's and policemen's pension systems have run up such massive deficits. One is that some systems have not operated on a reserve basis long enough to build up sufficient assets to make much of a dent in the liabilities. The Minneapolis systems are cases in point. In order to relieve the taxpayers' short-run burden, many systems have also operated on less than a fully funded basis. Under the charter amendment of 1959, Los Angeles was allowed to phase in its contributions in such a way that it did not have to start to fully amortize the unfunded liability for eleven years. Down through the 1950s, Detroit put money into the fire and police pension system to pay the bills for current service but not to cover the obligations for prior service. San Francisco followed a similar policy until 1979, when the retirement system's trustees prevailed on the Board of Supervisors to begin to amortize the unfunded liability. Under the Retirement Reform Act of 1979, Congress required only that the District of Columbia cap a portion of the unfunded liability over twenty-five years and thereafter pay the interest on it.[41]

Another reason many pension systems have run up such huge

deficits is that some cities made their contributions on the basis of what turned out to be erroneous actuarial assumptions. A good example is Oakland, which set up a funded system in 1951. On the advice of the actuary, the city put from 9 to 24 percent of payroll into the plan during its first twenty-five years. Unfortunately, the actuary underestimated disability rates, exaggerated withdrawal rates, and, worst of all, took no account of wage hikes, which averaged nearly 7 percent a year. Reviewing the situation in 1976, another actuary reported that the city should have been contributing 44 percent of payroll. To deal with the massive liabilities left by years of underfunding, he recommended that Oakland put in a staggering 129 percent of payroll—and raise employee contributions as well.[42]

Other cities ran into similar problems. After World War II, Detroit repeatedly underestimated fire and police salaries, to which pension benefits and postretirement adjustments were pegged. Between 1967 and 1976, salaries in Minneapolis went up more than two and a half times as fast as the rate on which the city based its contributions. From the 1940s to the 1970s, New York City firemen's and policemen's pension systems relied on actuarial assumptions that were developed for the Commission on Pensions in the 1910s. Based on a time when mortality rates were higher and retirement rates lower, these assumptions produced contribution rates that fell far short of meeting long-run obligations.[43]

In view of what actuaries do, it is small wonder that their assumptions are often wide of the mark. To figure out a pension system's costs, they have to estimate retirement age, death rates, wage increases, interest rates, and other variables. When it comes to death rates, actuaries can refer to many reliable mortality tables that have been developed over the years. But as far as the other variables go, they are pretty much in the dark. They have no choice but to extrapolate from prior experience, which, as a rule, leaves much to be desired. Nothing in the prior experience of the firemen's and policemen's pension plans foreshadowed the sharp climb in disability retirements, much less the great hike in salary levels, in the 1960s and 1970s.[44] To make matters worse, actuarial assumptions have profound political implications. Elected officials are well aware that a slight change in these assumptions—a rise, say, in the projected wage hike from 3 to 5 percent a year—will have a strong impact on recommended contributions. Pressed by the taxpayers to hold the line on expenditures, many officials bring pressure of their own on the actuaries to keep the contributions as low as possible. Private consultants in a competitive industry, whose estimates are at best educated guesses, actuaries are not always above shading their assumptions in a way that pleases their clients.

Still another reason many pension systems have run up such large deficits is that the authorities sometimes ignored the actuaries' recommendations or found other ways to reduce their contributions. The early history of Washington's Law Enforcement Officers and Fire Fighters System, which was set up in 1969, is instructive. In 1970, the actuaries recommended that the state put $47 million, or 33 percent of payroll, into the system. Caught in a fiscal squeeze, the legislature appropriated only $1 million. Three years later, the actuaries fixed the state's contribution at $63 million, or 38 percent of payroll. This time the legislature gave the system only $11 million. Combined with unexpectedly high wage hikes and disability rates, the state's tight-fisted policy drove the unfunded liabilities up from $240 million in 1970 to $612 million in 1975, or from 3.7 to 5.4 times payroll.[45]

The authorities cut their contributions in similar ways in Kansas City in the late 1950s and in Detroit in the early 1960s. In Los Angeles and New York City they came up with more ingenious gimmicks. In 1966, seven years after the Los Angeles Fire and Police Pension System went back on a reserve basis, the voters approved a charter amendment that gave the city seventy years, instead of forty-five, to amortize the unfunded liability. A year later, the New York State legislature changed the funding schedule for New York City's policemen's pension fund in a way that extended the amortization period from ten to thirty-five years, a step that lowered the city's contributions by 9 percent a year.[46]

Politicians often put too little money into the pension funds because it helps relieve pressure from taxpayers' groups, which want the authorities to hold the line on expenditures. It has the added advantage, a member of the Massachusetts Retirement Law Commission said in 1980, that the repercussions "will not usually be felt until the present group of public officials has left office."[47] Underfunding also helps relieve pressure from municipal employees, whose well-being depends on a steady flow of public funds, and from civic groups that insist on a high level of public services. Politicians justify their actions on the grounds that the city or state is caught up in the midst of a temporary financial crunch and that one year's reduction in contributions will not make much difference. But reduced contributions drive up unfunded liabilities, which in turn raise the costs of full funding and intensify pressures for underfunding. Willing to provide liberal retirement benefits but reluctant to put enough money into the funds and to make the uniformed officers contribute a fair share, politicians bear much of the responsibility for the morass in which most big-city firemen's and policemen's pension systems now find themselves.

CHAPTER FOUR

The Politics of Pension Reform

As the rolls swelled, the expenses soared, and the deficits climbed, the firemen's and policemen's pension systems aroused a good deal of concern among the upper middle and upper classes and the commercial and civic groups that spoke for them. This concern, which surfaced around World War I, built up in the 1920s and reached a peak during the Great Depression. The firemen's pension system, which had fallen "into the red" in 1926, was "again headed for disaster," the Portland City Club warned in 1932, and the policemen's fund was no better off. Drawing on a study that showed that the pension systems were "facing bankruptcy," the Minneapolis Taxpayers Association predicted in 1933 that "a breakdown is imminent."[1] The New York City Citizens Budget Commission, Chicago Civic Federation, and Los Angeles Municipal League issued similar warnings.

The unease over the firemen's and policemen's pension systems did not subside with the return of prosperity. The Philadelphia Bureau of Municipal Research, New Orleans Bureau of Governmental Research, Seattle Municipal League, and Los Angeles

Chamber of Commerce all called attention to the precarious condition of these plans in the 1950s. Their anxiety was shared by state pension or retirement commissions that had been established in Illinois, Minnesota, Wisconsin, Massachusetts, Washington, and New York after World War II.

Commercial and civic groups were troubled by the firemen's and policemen's pension systems because they posed a threat to the well-being of the cities' principal business interests—the banks, insurance companies, department stores, and real estate firms. The crux of the matter was that with some systems noncontributory and others only nominally contributory, the authorities had no way to pay the rapidly rising pension bills except by raising property taxes. As substantial property holders, the business interests were hit hard by what the Minneapolis Taxpayers Association described as a "phenomenal rise" in tax levies. Moreover, with most of these funds operating on a pay-as-you-go basis, pension costs would continue to go up in the future, as would property taxes. It would not be very long, warned Harold Riegelman, counsel to the Citizens Budget Commission, before the cities were forced to cut back on essential services in order to meet pension payrolls. To make matters worse, many cities did not even know the full extent of their obligations to the firemen and policemen—obligations that, as the Municipal League of Los Angeles put it, they had "*blindly* assumed" in the past.[2] Thus business interests were hard pressed not only to hold the line on taxes but also to figure out where that line would be.

Commercial and civic groups were also concerned about the firemen's and policemen's pension plans because they violated what many of the upper middle and upper classes regarded as the cardinal principles of good government. As the director of the University of Oregon's Bureau of Municipal Research and Service pointed out, the plans had developed in a "haphazard" manner, bearing no relationship to long-term personnel policies. Judged by the criteria of uniformity, equity, and efficiency, they left a good deal to be desired. Commercial and civic groups argued that the uniformed officers should not have separate pension systems. They also charged that the officers did not bear a fair share of the pension costs and that the cities did not put in enough money to meet the long-term obligations. Above all, commercial and civic groups were incensed by the eligibility requirements. It made no sense, they pointed out, to allow an officer to retire at any age after twenty or twenty-five years of service. Such a policy reduced contributions, raised expenditures, and, as the New Orleans Bureau of Governmental Research claimed, deprived the city of the services of veteran officers who were in "the prime of life." Nor did it make sense to give a pension to officers who were injured while off duty. While acknowledging the "special hazards" of policing and fire fighting, the Cleveland Citizens League still insisted that the officers should be protected against death and disability only if incurred in the line of duty.[3]

In an attempt to find a solution to the pension problem, commercial and civic groups sponsored a host of studies. Although some were done by their own members, others by actuaries, and still others by consulting firms, most of the studies reached similar conclusions. Based on these findings, commercial and civic groups urged that the firemen's and policemen's pension systems be consolidated with retirement plans for nonuniformed employees. If this were not feasible, the existing systems should be put on a reserve basis, with officers contributing their fair share. The eligibility requirements should be tightened as well: service requirements should be increased, age requirements imposed, and disability benefits restricted to officers injured in the line of duty. Finally, retirement allowances should be based on years of service—providing officers an incentive to remain on the job—and once granted, these allowances should not be pegged to changes in wages or to the cost of living.[4] The campaign to put these recommendations into effect, which was launched over sixty years ago, was supported by some elected officials and, after World War II, by most state retirement commissions.

The reform campaign made good headway in some cities. Under pressure from the Boston Finance Commission, the Massachusetts legislature closed the city's firemen's and policemen's pension funds to new appointees in 1923 and required that they join the less liberal city employees' retirement plan—a plan designed by George B. Buck and Paul Studensky. Eight years later, the Cincinnati City Council adopted a similar plan, also prepared by Buck under the auspices of Cincinnati's Bureau of Municipal Research. Under this plan, the firemen's and policemen's pension funds were closed, and new recruits were enrolled in the city employees' retirement system, a funded system with stringent eligibility requirements and modest pension benefits. Shortly after World War II, the Wisconsin legislature set up a consolidated pension system for Milwaukee's firemen, policemen, and nonuniformed employees, a scheme favored by the Milwaukee City Club. The new system, which applied only to firemen and policemen appointed after the law went into effect, established stiffer eligibility requirements, lower retirement benefits, and a higher contribution rate.[5]

But the reform campaign ran into heavy resistance in other cities. A case in point is Los Angeles, where the City Council lowered the age requirement from sixty to fifty-five in 1913 and abolished it altogether six years later. During the mid-1920s, the Municipal League, observing that many able-bodied officers were retiring in "the prime of life," called on the council to impose an age requirement of fifty. The Board of Pension Commissioners endorsed the proposal, but the Fire and Police Protective League objected so strongly that the council refused to submit it to the voters. Several years later, in the midst of the financial crunch that led the city to place the pension system back on a pay-as-you-go basis, the Los Angeles Realty Board urged the council to put a charter amendment on the ballot setting an age requirement of sixty. Although the *Los Angeles Times* favored the board's position, the council refused to go along. Yet another reform effort got under way in the mid-1950s, when a citizens committee appointed by Mayor Norris Poulson issued a report recommending that the city impose an age requirement of fifty-five and raise the service requirement from twenty to twenty-five years. The Protective League sharply opposed the recommendations. Over the objections of the mayor, the *Times*, and the Chamber of Commerce, which wanted the proposed change incorporated into a charter amendment and placed on the ballot, the council shelved the report.[6]

Nor was Los Angeles atypical. New York City's firemen and policemen fought off an attempt to move them into the city employees' retirement system in the early 1920s and thwarted an effort to put them into a statewide pension plan in the early 1930s. They later stymied attempts by the Citizens Budget Commission to raise contribution rates, tighten eligibility requirements, reduce retirement benefits, and put their systems on a reserve basis. Firemen and policemen also headed off pension reform in Detroit and New Orleans. Prodded by Detroit's Bureau of Governmental Research, the voters established a Pension Commission in 1933. Chaired by A. G. Gabriel, a prominent actuary, the commission recommended that the city set up a consolidated retirement system for all its employees. Aware that the requirements would be stiffer and the benefits lower under the proposed plan, the uniformed forces lobbied hard and eventually prevailed on the City Council to exclude them from the new system, which was adopted in 1937. The reform effort in New Orleans was launched in the mid-1940s by the Bureau of Governmental Research, which was alarmed by soaring pension costs. At its urging, the Louisiana legislature eliminated a provision that allowed policemen to retire on 40 percent of salary after sixteen years of service. But the legislature balked at the bureau's proposal to put new employees into a consolidated retirement plan, a plan that would have imposed an age requirement of sixty, based benefits on contributions, and operated on a funded basis.[7]

The reform efforts may have eased the pension problem, but they did not solve it. They failed not for lack of concern or hard work, nor for lack of solutions. They failed rather because the constraints on pension reform were too strong. To identify these constraints, and to fully appreciate their strength, it is useful to look closely at a few case studies of pension reform.

The first case study deals with a reform proposal that was put forth in Los Angeles in the 1960s but ran into so much opposition that the City Council refused to go along with it. The second is about a reform measure that was adopted in New York City in 1940, subjected to sharp criticism afterward, and undone a decade later. The third deals with a reform proposal that was approved by the Washington legislature in the 1960s, but only after it was completely watered down. Each of these studies depicts a different way in which pension reform was thwarted. The fourth case describes how Detroit managed to tighten the eligibility requirements and put a ceiling on the cost-of-living escalator in the 1960s. Although a success, the Detroit effort reveals a good deal about the politics of pension reform. Along with the three failures, it provides a basis on which to draw some generalizations about the political constraints on pension reform in urban America.

The Los Angeles case goes back to the early 1960s. Shortly after the voters put the Los Angeles Fire and Police Pension System (LAFPPS) back on a reserve basis, pressure began to build to reform the system again. The pressure came mainly from the Chamber of Commerce, spokesman for the city's commercial and financial elite and successor to the Municipal League as the pension system's watchdog. The chamber was deeply concerned about rising pension costs. To take the LAFPPS off a pay-as-you-go basis was not enough, the chamber insisted; it was also imperative to cut the system's costs, either by tightening eligibility requirements or by reducing retirement benefits. Along with several taxpayers groups, the chamber had been pushing for such changes for well over a decade; but it had been stymied by the Los Angeles Fire and Police Protective League. The league, spokesman for the uniformed forces since the 1920s, was also troubled, but by a different problem. Many retired officers and their widows were on a fixed pension that was based on the meager salaries of the 1930s and 1940s. With a steady rise in the cost of living, some of them were close to destitute. Although the voters had turned down a charter amendment in 1959 that would have provided a cost-of-living adjustment for disabled firemen and policemen, the Protective League was nonetheless determined to seek for its members some sort of hedge against inflation in the years ahead.[8]

The City Council responded to these pressures by instructing City Administrative Officer C. Erwin Piper to do a study of the LAFPPS and figure out how to revise the system. A former Federal Bureau of Investigation agent, with a Ph.D. in public administration and a profound grasp of local politics, Piper was probably as well qualified for this task as anyone in Los Angeles. Starting in late 1963, he and his staff held a series of separate meetings with each of the groups that was vitally interested in the policies of the LAFPPS. The meetings dealt with eligibility requirements, retirement benefits, contribution rates, funding arrangements, and cost-of-living adjustments.

Two contrasting viewpoints were spelled out in the meetings. One, which was expressed by the Chamber of Commerce, California Taxpayers Association, and Property Owners Tax Association of California, held that the pension system was far too liberal. Its advocates recommended that the city reimpose an age requirement, raise the service requirement, lower the disability benefits, and retain a fixed pension. The other point of view, which was articulated by the Protective League, Police Chief William H. Parker, and a few retired officers' groups, held that the system was not liberal enough. Its supporters proposed that the city not only maintain the current eligibility requirements and retirement benefits but also establish a postretirement escalator pegged either to fire and police salaries or to the cost of living.[9]

Following the meetings, Piper and his staff drafted a new pension plan that they claimed would reduce the taxpayers' burden without lowering the officers' benefits. In November 1965, Piper submitted his recommendations to the City Council. Under the proposed system, which would cover new entrants and veteran officers who elected to join it, firemen and policemen would have to serve until fifty and put in twenty-five years, instead of twenty, to be eligible for a service pension. At fifty-five (or, for disabled officers, after five years on a pension, whichever came first), the retired officers or their survivors would receive an automatic 1.5 percent a year increase in their base pension. To pay for the new system, members would contribute 6 percent of their salaries, plus one-half the costs of the cost-of-living adjustment, and the city would put in enough money to cover the long-term costs and the interest on the unfunded liability. To alleviate the plight of retirees whose benefits were tied to the salary levels of the 1930s and 1940s, the minimum pension would be $150 a month if the recipient were single and $200 a month if he were married.[10] All things considered, Piper's scheme was an ingenious effort to provide relief for the taxpayers and security for the officers in a way that it was thought would be acceptable both to the Chamber of Commerce and to the Protective League.

But the scheme was not acceptable to the Protective League, whose president labeled it "a slap in the face to every policeman and fireman." At the core of the league's opposition were the proposed changes in the eligibility requirements. The imposition of an age requirement was misguided, the league insisted, because policing and fire fighting were extremely hazardous, quasi-military occupations for which only young men in top-notch physical condition were suited. An extension of the service requirement was inadvisable at a time when the fire and police departments were finding it hard to attract qualified recruits. To slow the growth of the pension rolls, the city should not tighten eligibility requirements, the league argued; instead it should offer veteran officers incentives to stay on the job. It should tie retirement allowances to years of service, giving officers a higher percentage of final salary after twenty-five years than after twenty and an even higher percentage after thirty years. After retirement, allowances should be pegged to the cost of living, the league contended. To hold down costs, adjustments should be restricted to service pensioners, deferred until the twenty-fifth anniversary of their appointments, and limited to 2 percent per year.[11] The league incorporated these recommendations into a plan that was submitted to the City Council in June 1966.

Convinced that the plan stood a better chance at the general election in the fall than at a municipal election in the spring, when the turnout would be low, the league urged the council to put its proposal on the November ballot. Its spokesmen stressed that the plan would not only protect retired officers against inflation and help the departments attract qualified recruits but also reduce pension costs by encouraging veteran officers to stay on the job. Piper, who thought that the LAFPPS was already much too liberal, spoke out against the league's proposal. It offered the taxpayers nothing in return for the cost-of-living adjustment and would cost an additional $5 million a year. The city would do better to adopt a modified version of his plan, which retained a service requirement of twenty years, provided a 2 percent a year cost-of-living adjustment, and raised the minimum allowance to $250 a month for retired officers and to $200 for their widows. Also opposed to the league's plan was the Chamber of Commerce. Its spokesmen objected on the grounds not only that the plan would cost an additional $5 million a year but that it would allow officers to retire in their early forties. By requiring the city to pay only the interest on the unfunded liability, it would also weaken the funding arrangements approved in 1959. Along with the California Taxpayers Association, the chamber urged the council to defer action on the Protective League's request.[12]

But, as Piper put it a decade later, the council was "just going through the motions." The outcome was never in doubt. The league, which, in his words, had "more political clout than any other group in city government," left nothing to chance. It got endorsements from the police chief, fire chief, and members of the fire and police commissions, thereby demonstrating that labor and management were united. It also lined up the support of the Board of Pension Commissioners. In the meantime, the league's leaders did extensive lobbying in the City Council, with many of whose members they enjoyed close personal and working relationships. The councilmen, for their part, knew that the uniformed officers and their families made up a large bloc of voters. They also remembered that ten years earlier hundreds of off-duty firemen and policemen had campaigned door to door on behalf of a proposal to give the legislators what they regarded as a well-deserved raise. And in the aftermath of the Watts riots of August 1965, which devastated the south-central ghetto and terrified the rest of Los Angeles, the councilmen were reluctant to take any action that implied less than full support for the uniformed forces. So after making some minor revisions—notably one that required the city to continue to amortize the unfunded liability but that extended the schedule from forty-five to seventy years—the council voted to submit a charter amendment embodying the league's plan to the voters.[13]

The league then launched a campaign on behalf of its plan, which was listed on the ballot as Proposition P. It persuaded a group of eminent citizens, headed by former Governor Goodwin Knight, to come out in favor of the proposition. It also got endorsements from the *Times* and *Examiner*, the mayor, ten of the fifteen councilmen, and fourteen of the fifteen members of the fire, police, and pension commissions. Meanwhile, off-duty officers prepared mailings in fire stations, which operated much like ward clubs. They also went door to door canvassing and, under the supervision of two speakers bureaus, gave speeches to civic groups and service clubs. The league purchased about $100,000 worth of newspaper, radio, and television ads as well. To pay for them, most officers asked the city controller to deduct 1 percent from their paychecks in the months preceding the election. The league stressed the same points it made to the council in August. Besides modernizing an outdated system, Proposition P would help attract qualified personnel and retain veteran officers, thereby enhancing public safety—an argument that struck a responsive chord in the wake of the Watts riots. It "will not require an increase in taxes!" either. Speaking for the opposition, which included the Chamber of Commerce and the Realty Board, the Property Owners Tax Association of California warned that Proposition P would raise taxes by $5 million per year. But its warnings carried little weight with the voters, who approved the league's plan by a majority of slightly more than three to two.[14]

Thwarted by the firemen and policemen, the reform effort was a dismal failure. And the worst was yet to come. As a concession to the Chamber of Commerce and to taxpayers groups, the league had limited the cost-of-living escalator to service pensioners and had imposed a 2 percent ceiling on it. But not long after Proposition P was approved, the league moved to get rid of these restrictions. In 1969, it brought pressure on the City Council to put on the ballot a charter amendment extending the escalator to disability pensioners. The council gave in. Although the Chamber of Commerce and taxpayers groups were far from enthusiastic about the proposed change, they could not come up with a compelling reason why the city should give an escalator to some retirees and not to others. The voters adopted the amendment by nearly two to one. Two years later, the league drafted another charter amendment lifting the ceiling on the escalator and pegging it to the consumer price index (CPI). The Chamber of Commerce and taxpayers groups were dismayed by the scheme, the fiscal implications of which were mind-boggling, but they were hard pressed to explain why the escalator should be limited to 2 percent. In yet another demonstration of its political clout, the league prevailed on the City Council to put the proposal on the ballot. After a vigorous campaign, in which the firemen and policemen contributed more than $300,000 to the league's war chest, the amendment passed by a majority of less than 2 percent.[15] The CPI rose so much in the 1970s that by the end of the decade the cost-of-living escalator cost the taxpayers more than all the system's other benefits combined.

The firemen and policemen opposed pension reform in New York City as well, but for a decade it looked as if they would have to live with it. And only after a lengthy struggle did they prevail on the authorities to undo the reform effort. The story begins late in the 1930s, when Mayor Fiorello H. LaGuardia, fresh from a resounding victory in his bid for reelection, launched another in a long series of attempts to reform the city's firemen's and policemen's pension systems. At the time, the policemen's fund, to which the members contributed 2 percent of salary, provided half-pay after twenty-five years; the firemen's fund, which was noncontributory, gave half-pay after twenty years. Operating on a pay-as-you-go basis, the funds were running massive deficits, which, some argued, would soon force the city to put off capital improvements, cut back vital services, and lay off essential personnel.[16]

Well aware that the uniformed forces had blocked previous reform efforts, LaGuardia sought the cooperation of the heads of the rank-and-file associations. They were far from enthusiastic about pension reform. But fearful that if they refused to cooperate the mayor might throw his weight behind a tough bill that had been introduced into the state legislature on behalf of the Citizens Budget Commission and other civic groups, they helped design a new pension plan. LaGuardia made it public in November 1939. "Pension Reform at Last," declared the *New York Times*, fully two decades after the city's Commission on Pensions first called attention to the problem. As part of the understanding with LaGuardia, the firemen and policemen held a referendum shortly after; and under enormous pressure from their leaders, a sizable majority endorsed the mayor's proposal.[17]

Under the mayor's proposal, there would be four separate plans for firemen and policemen. Current officers would remain in the old pay-as-you-go plans, both of which would be known as Article I. They could retire after twenty years by contributing 6 percent of salary or after twenty-five years by putting in 5 percent. Their death, disability, and survivors benefits would be unchanged. Future entrants would have to join two new funded plans (Article II for policemen and Article IA for firemen) and pay 45 percent of the costs. While they too could retire after twenty years, it would cost them up to 10 percent of salary—and up to 18 percent if they wanted the benefits passed on to their widows.

The Patrolmen's Benevolent Association (PBA) and other rank-and-file groups, all of whose members were covered by the more liberal Article I plans, favored the proposal. So did the Citizens Budget Commission. Although the commission would have preferred to see an age requirement included in the new systems, it realized that few officers would elect to retire after twenty years at the new contribution rates. It also believed that if officers were required to pay 45 percent of the costs, they would be reluctant to press for improved benefits in the future.

What little opposition there was came largely from the Patrolmen's Eligibles Association, which represented about 1,400 candidates who had been certified by the Civil Service Commission but not yet appointed to the force. The association objected to the mayor's proposal on the grounds that their members, who would do the same work as other officers, would have to contribute three times as much to get the same benefits. Despite the association's objections, the bills incorporating the proposal were passed by the City Council, approved by the Board of Estimate, and signed by the mayor early in 1940.[18]

The opposition to the new system built up shortly after it went into effect in April 1940. Early in 1941, the Uniformed Firemen's Association (UFA) complained to George B. Buck, the city's actuary, that Article IA contribution rates were too high. Buck responded that the city could lower the rates if the firemen would accept slightly reduced benefits. At the UFA's request, Buck drafted another article, known as IB, which cut the rates 2 to 3 percentage points. The mayor and the City Council went along with it, and in July 1941 all but one member of Article IA transferred into Article IB. The PBA wanted to sponsor similar legislation, but at the urging of the Article II officers, whose spokesmen claimed that such a change would only make things worse, it held off. Not long after, Article II policemen, with the blessings of PBA leaders, formed a group to look into the prospects of revising their pension system.[19] It was this group, known as Pension Forum, Inc., that led the campaign to undo pension reform in New York City.

At the outset, Pension Forum's prospects were bleak. A small group with little money and power of its own, it had very little influence in the PBA. The Article II members for whom it spoke made up a small minority of the rank and file, roughly 1,800 out of more than 15,600 in 1943. Moreover, only a handful of them sat in the House of Delegates, the association's governing body. To make matters worse, relations between Pension Forum and the PBA were far from cordial. Pension Forum charged that the PBA's leaders were paying little attention to the pension problem and otherwise ignoring the concerns of Article II officers. PBA leaders countered that Pension Forum was taking a parochial position that was destroying the unity of the association and undermining the interests of the rank and file.[20]

Even if Pension Forum had prevailed on the PBA to take a more aggressive position on pensions, it probably would not have made much difference. High contribution rates for policemen would not have been a compelling issue in the middle of World War II. The Citizens Budget Commission and other civic groups would have staunchly resisted any changes in the new system. And they could probably have counted on LaGuardia, who was still in power, to head off any challenges to his settlement. Thus Pension Forum kept the issue alive through the mid-1940s, but only barely.

The situation began to change after World War II. With many recruits joining the force, the number of Article II members rose to roughly 6,000, which was a large chunk of the PBA. At the same time, their leaders picked up additional seats in the House of Delegates, giving them more of a voice in the association's policies, including its position on pensions. Relations between the PBA and Pension Forum still remained strained; at one point the association expelled a few Pension Forum members from the House of Delegates, only to see them reinstated later by the courts. But PBA leaders, caught up in a struggle for control, could no longer dismiss the demands of the Article II members. With the strong support of the UFA, the PBA finally mounted a campaign to revise the pension plans in the late 1940s. Its demands were incorporated into two bills that were introduced into the City Council in 1949 and debated in 1950. One would have lowered the contribution rates by reducing the employees' share of the costs from 45 to 25 percent and raising the city's share from 55 to 75 percent. The other would have permitted the policemen to transfer from Article II to Article I. As expected, the Citizens Budget Commission, Board of Trade, and other civic and commercial groups lashed out against the bills on the ground that the costs would be prohibitive.[21]

Despite the opposition, the PBA was optimistic. After a dozen years as mayor, LaGuardia had retired in 1945. His successor, William O'Dwyer, a former policeman and Tammany stalwart, was running for reelection in 1949. With Tammany in disarray, the mayor sorely needed the support of the policemen, firemen, and other municipal employees. To get it, he promised to do his utmost to cut the contribution rates. O'Dwyer was reelected in November but resigned under fire nine months later. He was replaced by City Council President Vincent Impellitteri, who decided to run as an Independent to fill the remaining years of O'Dwyer's term. When Impellitteri refused to promise to support a change in the pension systems, the PBA threw its weight behind Tammany's candidate, Ferdinand Pecora. In return, Tammany agreed to back the association's campaign.[22]

The issue came to a head in the City Council in October 1950. The PBA insisted that the contribution rates were so high that the rank and file could not make ends meet. Speaking for the upper middle and upper classes, the *Times* took the position that if take-home pay was too low, the council should raise salaries, not lower contributions. The Citizens Budget Commission denounced the legislation as "inequitable, untimely, financially onerous and of doubtful legality." Despite these objections, the council voted to reduce the officers' share of the expenses from 45 to 25 percent—although it did not go along with the proposal to allow policemen to transfer from Article II to Article I.[23]

The legislation then came before the Board of Estimate, which was chaired by the mayor. The board decided to defer a decision until after the election; and when Impellitteri won an upset victory, it tabled the measure. Infuriated, the policemen and firemen held a rally at City Hall and called on organized labor for support. But the campaign seemed doomed. The legislation "is dead," Pension Forum announced, putting the blame on the ineptitude of the PBA leadership. The legislation did not stay dead long, however, because in February 1951, Michael Quill, head of the Transport Workers Union (TWU), announced his intention to organize the police force. His efforts attracted a good deal of interest among the rank and file. But it dismayed the upper middle and upper classes, who were frightened by the prospect of police unionism, especially under as militant a unionist as Quill. Nor were they reassured by Commissioner George Monaghan's statement that under no circumstances would he recognize the TWU as the bargaining agent for the rank and file. John Carton, president of the PBA, skillfully exploited these fears, warning elected officials, "If you don't want policemen delivered to Quill on a platter, pass the pension bills." With Quill pressing ahead, many New Yorkers concluded that perhaps the city had ignored the "just demands" of the police force; a revised pension system might be bad, they decided, but a police union would be even worse.[24]

Responding to this turnabout, the Board of Estimate reversed itself in early August; it voted unanimously to reduce the employees' contributions and referred the pension bills back to the City Council. On the same day, Commissioner Monaghan undercut the TWU's organizing drive by forbidding the rank and file to belong to a labor union. A week later, the City Council, with but one dissenting vote, passed the bill, again cutting the employees' share of the costs from 45 to 25 percent and raising the city's share from 55 to 75 percent. This time the Board of Estimate went along. And despite opposition from the Citizens Budget Commission and the *Times*, which dubbed the bills "too costly a method of defeating Mr. Quill's effort to unionize the police," Impellitteri gave his approval. The legislation did not allow the officers to transfer from the new systems to the old ones; nor did it take the new systems off a reserve basis. But it reduced the firemen's contribution rate by 5 to 8 percentage points and the policemen's by 7 to 10 points, thereby increasing the city's burden by several million dollars a year. By reducing the officers' share of the costs from nearly one-half to only one-quarter, the legislation also gave the firemen and policemen a strong incentive to push for improved benefits. And in the years ahead they did just that.[25] For the time being, however, the uniformed officers were happy to leave well enough alone, satisfied that a decade after the establishment of Article II and Article IA they had managed to undo pension reform in New York City.

The firemen and policemen stymied pension reform in the state of Washington in a different way. They blocked several reform proposals and then supported one that had been watered down to their satisfaction in the legislative process. The story begins in the early 1960s, when the Washington legislature, under pressure from the

Seattle Municipal League and the Association of Washington Cities to do something about rising pension costs, ordered a study of public retirement plans in the state. The study was delegated to a subcommittee of the Legislative Council, a bipartisan body of senators and representatives, which retained the services of A. A. Weinberg, the prominent public sector actuary. Weinberg's report, which came out in April 1962, called attention to a good many problems, one of the most serious of which was the financial condition of the local firemen's and policemen's pension plans. The plans were quite liberal, although by no means atypical. Besides providing adequate disability benefits, they allowed the officers to retire at half-pay after twenty-five years of service—policemen at any age, firemen at fifty. The policemen's plans also granted post-retirement increases, which were pegged to active salaries. For this, the officers contributed 6 percent of salary. Operating more or less on a pay-as-you-go basis, the systems had run up sizable deficits—$47 million ($29,000 for each officer) in the policemen's funds and $58 million ($24,000 per officer) in the firemen's funds.[26] With the two services vying to outdo one another in prying more generous benefits out of the legislature, the system's long-term solvency was very much in doubt.

To resolve the problem the Legislative Council recommended the formation of a retirement plan for all firemen and policemen that would be managed by the existing statewide or citywide systems for municipal employees. (The council opposed the formation of an independent statewide public safety officers system, an alternative put forth by the Association of Washington Cities, on the grounds that policing and fire fighting were local responsibilities.) The plan, which was drafted by Weinberg, covered new entrants (and, if they wished, current officers). It lowered the service requirement to twenty years but imposed an age requirement of fifty-five. It also raised the maximum benefits from 50 to 75 percent of final salary but redefined final salary in a way that reduced retirement allowances. It abolished the policemen's cost-of-living escalator as well. To cover the costs, Weinberg proposed raising employee contributions from 6 to 8 percent and requiring the cities to put in 14 percent of payroll, plus enough to amortize the prior service liabilities over thirty years.[27]

The big-city policemen and firemen strongly objected to the council's proposal, arguing that it would be preferable to maintain the existing plans and, if need be, to put them on a reserve basis. The municipal officials, whatever their reservations about the principle of pay-as-you-go, were far from enthusiastic about shelling out a minimum of 14 percent of payroll. Thus, while the council's proposal was incorporated into several bills and submitted to the 1963 legislative session, it made no headway in the face of so much opposition.[28]

Prospects for pension reform picked up in 1965, when the state legislature bowed to pressure from the Legislative Council and Association of Washington Cities and set up a Public Pension Commission. One of several such commissions formed after World War II, it was a bipartisan body made up of eleven legislators and four distinguished citizens. Chaired by John Ryder, a Seattle Republican and minority leader of the Senate, and later by Homer Humiston, a Tacoma Republican and chairman of the House Committee on Local Government, the commission quickly emerged as the principal advocate of pension reform in the state. Drawing heavily on the Legislative Council's study, the commission introduced a bill of its own, known as H.B. 113, in July 1966. The bill called for a statewide retirement plan for all firemen and policemen—a funded plan that would be administered by the state and financed by the cities and that would cover all officers appointed on or after January 1, 1968. Current employees would remain in the existing systems. The commission kept the service requirement at twenty-five years, set the employees' contributions at 6 percent, and fixed the cities' contributions at 18 percent. Otherwise the bill stuck closely to Weinberg's plan. Speaking in favor of the commission's bill, Ryder argued that it would put the firemen's and policemen's pensions on a sound fiscal footing and ease the crushing financial burden on the municipalities.[29]

The Seattle Municipal League supported the bill, taking the position that while the proposed plan was too liberal it was an improvement over the current system. Thomas P. Bleakney, a well-known actuary with Milliman & Robertson, endorsed it, too. But the Association of Washington Cities opposed the bill, insisting that the cities could not afford to put 18 percent of payroll into the system. Speaking for the association, Mayor Harold Tollefson of Tacoma told the House Committee on Local Government, "It is just not within our budgets to absorb it [the 18 percent] without cutting other services." The firemen and policemen also opposed the bill. According to their spokesmen, the existence of two different systems, one for current employees and another for new entrants, would be a source of friction in the years ahead. They attacked the proposed age requirement of fifty-five on the grounds that policing and fire fighting were extremely hazardous occupations that needed young men in prime physical condition. They also criticized the redefinition of final salary and the elimination of the escalator clause on the grounds that a reduction in benefits would make it much harder to attract qualified recruits.[30] With the municipal officials and uniformed services solidly opposed, the Pension Commission's bill died in committee in 1967.

Following the demise of H.B. 113, the Pension Commission, then chaired by Humiston, drafted another bill, known as H.B. 950, which was submitted to a special session of the legislature and referred to the House Committee on Local Government. Designed to overcome the objections of the uniformed forces and municipal officials, H.B. 950 differed in several ways from H.B. 113. The new bill permitted firemen and policemen who had put in twenty-five years to retire before reaching fifty-five, although on a smaller pension. Another difference was that applications for disability pensions were to be processed by local boards, on which the officers would be well represented, rather than by the state Public Employees Retirement System, which would administer the plan. The new bill also defined final salary as average salary in the officer's two best rather than five best years (as it was defined in H.B. 113). Lastly, H.B. 950 provided that all retired officers who were at least fifty-five would receive a cost-of-living increase of 1.5 percent a year for the remainder of their lives. Under H.B. 950 the officers would contribute 7 percent of salary, 1 percent of which was for the proposed escalator. As a concession to municipal officials, the cities would put only 7 percent of payroll into the system; the rest of the costs, which meant the bulk of them, would be picked up by the state.[31]

The Association of Washington Cities had a few reservations about the bill. The officers would not put in enough money, it said; a contribution rate of 8 or 9 percent would be preferable. Nor was there any reason for the plan to provide non-service-connected disability pensions. But if the state was willing to pay the bills, the association would not stand in the way. Bleakney, who had long favored replacing the local pension systems with a statewide retirement plan, endorsed the bill. And Chief Gordon F. Vickery of the Seattle Fire Department, who had vigorously opposed H.B. 113, went on record in favor of H.B. 950, an action for which he was condemned by Local 27 of the International Association of Fire Fighters.[32]

Humiston had promised Charles Marsh and John Smith, Jr., the principal lobbyists for the policemen and firemen, respectively, that he would not let the bill out of committee unless the uniformed forces supported it. Despite the concessions, they did not. The firemen and policemen took strong exception to the imposition of an age requirement of fifty-five, the redefinition of final salary, and the modification of the cost-of-living escalator. Rather than set up a new plan, their spokesmen argued, the legislature should give financial aid to the old systems. Honoring his commitment to the uniformed forces, Humiston refused to hold public hearings on the bill and kept it bottled up in committee.[33]

Undaunted, the Pension Commission drafted yet another bill, known as H.B. 353. Introduced into the legislature early in 1969, it called for a statewide Law Enforcement Officers and Fire Fighters Retirement System (LEOFF). H.B. 353 was backed by the Association of Washington Cities, whose members saw it as a way to escape rising pension costs. It was also supported by the small-town firemen and policemen, many of whom lacked an adequate pension plan. Designed to overcome the opposition of the big-city officers, LEOFF was fairly liberal to begin with. As amended by the House Labor Committee, it was among the most liberal systems in the country. LEOFF permitted retirement at half-pay at fifty after twenty-five years of service. It eased the eligibility requirements for disability pensions, the applications for which would be processed by local boards. It left the definition of final salary unchanged and provided a cost-of-living escalator pegged to the CPI. LEOFF also guaranteed current officers at least as much in benefits as they would have gotten under the old plans. And it required the cities to pick up the retired officers' medical bills for the rest of their lives. For all this, firemen and policemen would contribute 6 percent of salary and the cities another 6 percent; the state would not only pay the rest but also assume the unfunded liabilities. With nothing to lose, the uniformed personnel gave their approval to H.B. 353, which sailed through the legislature in 1969 and went into effect a year later.[34]

Thus pension reform came to Washington, but at a high price. Just how high was soon revealed. Under LEOFF, the disability rolls went up at a rate that "staggers the imagination," to quote Bleakney, the system's actuary. With the cost-of-living escalator pegged to the CPI, which reached double digits in the early 1970s, the retirement allowances soared as well. And as the state, pleading budgetary problems, put in only a small fraction of the recommended contributions, the unfunded liabilities climbed from about $240 million in 1970 to $612 million in 1975. Not long after LEOFF went into effect, Senator Martin J. Durkan conceded that "the state was sold a bill of goods." Representative Helen Sommers called LEOFF the state's "most serious financial problem," a view shared by Joseph R. Curtis, president of the Washington State Research Council. Jack Rogers, executive secretary of the Association of Washington Cities, which had lobbied hard on behalf of H.B. 353, declared that LEOFF was "near disaster."[35] Many legislators realized that in an effort to overcome the opposition of the big-city firemen and policemen to pension reform, they had made a liberal retirement system even more liberal. Instead of solving the pension problem, they had simply shifted the burden from the cities to the state.

The Detroit case—the one successful example of pension reform included here—got under way in the mid-1950s, at which time the City Council was deeply concerned about the financial condition of the firemen's and policemen's pension system. Its costs were going up so fast that in 1957 the city had to put $11.2 million, or 33 percent of payroll, into the fund—up from $1.5 million, or 10 percent of payroll, in 1947. Although the system operated on a reserve basis, its unfunded liabilities soared from $49 million in 1947 to $120 million in 1957, roughly $20,000 for each officer and nearly 3.6 times payroll. Unsure what to do, the City Council appointed two committees to look into the problem. Both reported that the pension system was heading for serious trouble, a finding confirmed by the Public Service Administration, a Chicago consulting firm that made a survey of several big-city firemen's and policemen's pension funds in the early 1960s. The crux of Detroit's pension problem was that the system allowed the officers to retire too early, thereby reducing revenues and raising expenses, and provided them a cost-of-living escalator that was pegged to fire and police salaries. To resolve the problem the studies recommended that the city impose an age requirement, put a ceiling on the escalator, and link retirement allowances to length of service—a scheme that would give the officers an incentive to remain on the job after becoming eligible for retirement.[36]

The City Council took no action on these recommendations, thus bequeathing the problem to Mayor Jerome Cavanagh, who was plagued by it for most of his two terms of office. Elected in 1961, Cavanagh inherited a deficit of $35 million, 45 percent of which was owed to Detroit's pension funds. The debt was incurred in 1961, when the courts ruled that the council had put too little money into the systems in the past and ordered it to start making back payments. To meet the costs, the Cavanagh administration enacted a city income tax in 1962; and three years later, the city paid the last installments to the pension systems. No sooner had Cavanagh wiped out the debt than Commissioner Ray Girardin declared that the police department was facing a grave manpower shortage. Citing early retirement as the main problem, he urged the city to offer incentives to stay on the job to officers who had served twenty-five years. Stop deducting 5 percent from their paychecks for pensions, he suggested, or raise their allowances by 2 percent for each additional year of service. To make matters worse, the pension payroll was going up so rapidly that the city had to put more than $18 million into the pension system in 1967—an increase of more than $7 million since 1957. Despite the enormous contributions, which came to about 41 percent of payroll, the unfunded liabilities that year came to $188 million—more than $31,000 for each officer and nearly 4.2 times payroll.[37]

A series of seemingly unrelated events brought the issue to a head in 1968. Two years earlier, the Detroit Police Officers Association (DPOA) was recognized as the bargaining agent for the rank and file. Shortly after, it started negotiating with the city. The negotiations went slowly, partly because both sides were uncertain about the ground rules and partly because the city was strapped for funds. As the talks dragged on, the DPOA grew increasingly frustrated and at one point even filed charges of unfair labor practices against the city. The Cavanagh administration gave the policemen a $1,000 raise in 1966, which reduced the tension a bit. But when it held the line on wages a year later, the officers resorted to direct action. Under the aegis of the DPOA, they stopped writing tickets and then called in sick, setting off the "blue flu" epidemic of June 1967. In an effort to get the policemen back to work, the administration agreed to submit the unresolved issues to a fact-finding panel. Before the panel was formed, however, the worst rioting in decades broke out in Detroit, leaving forty-three dead, four hundred injured, and $50 million in property damage. The riot had a strong impact on the panel, which issued its report in February 1968. Emphasizing the department's manpower problems and downplaying the city's fiscal troubles, it endorsed the DPOA's demand for a base salary of $10,000.[38]

Concerned about the prospect of further rioting and thus reluctant to antagonize the police force, Cavanagh went along with the panel's recommendation. With only slight modification, it was soon put in effect, boosting the policemen's base pay by $1,665. As was customary in Detroit, the firemen received the same raise. More alarming to the mayor than the wage hikes was their impact on pension costs. With both the retirement allowances and the cost-of-living escalator tied to fire and police salaries, it was estimated that the city's contributions to the pension system would double in three years—climbing to $35 million in 1971, or a whopping 52 percent of payroll. After informing the DPOA of his fears, Cavanagh included in his annual budget message to the City Council a far-reaching proposal for pension reform. Drawing on a report by a task force on city finances, which relied heavily on the two studies done in the 1950s, the mayor suggested that the city impose an age requirement of fifty-five. To provide veteran officers an incentive to stay on the job after fifty-five, he recommended that retirement allowances be based on years of service and that they be vested after twenty-five years on the job. Cavanagh also proposed that the city stop tying retired officers' allowances to fire and police salaries and instead give them an annual 2 percent cost-of-living adjustment. The mayor urged the council to incorporate these recommendations into a charter amendment and submit it to the voters at the fall election.[39]

The policemen and firemen, represented by the DPOA, Detroit Firefighters Association (DFA), and a few superior officers' groups, strongly objected to Cavanagh's proposal. Stressing that under the city charter only the voters could revise the pension system, the city's negotiators offered to talk with the unions about the proposed changes—but only about these changes. Spokesmen for the unions replied that they would only discuss pensions together with work rules and other issues that were in dispute. After several months of fruitless negotiations the showdown came in September 1968. At a City Council meeting packed with angry firemen and policemen, the unions' expert insisted that the reforms were unnecessary; the city could cut costs by one-third simply by updating the mortality tables and revising the investment policies. The city's actuary responded that any savings from these changes would be offset by the system's policy of deliberately underestimating future salaries of firemen and policemen. Although the City Council normally deferred to the wishes of the uniformed services, it was under intense pressure to do something about the rising costs of the pension system, which one legislator called "an albatross around the city's neck." Looking at the mayor's proposal, whatever its shortcomings, as a "stopgap measure" that might hold down pension costs and at any rate would take some of the pressure off, the council voted unanimously to put the amendment on the November ballot.[40]

If the council put the amendment on the ballot, DFA president Earl J. Berry vowed a week or so before the meeting, "We'll bust open our treasuries to fight it."[41] Carl Parsell, head of the DPOA, felt the same way. Starting in September, the DFA deducted an extra $5 a month from each fireman's paycheck to finance the campaign against the proposal. The DPOA assessed each patrolman an additional $10 a month; and the sergeants', lieutenants', and detectives' associations raised their dues $2 a month.

To coordinate the campaign, the unions formed the Citizens Committee for Responsible Detroit Government. The committee, which included Berry, Parsell, and several local labor leaders, placed ads on radio and television and bought space in newspapers and on billboards. Under its supervision, the rank and file handed out leaflets and went canvassing door to door. The policemen also put bumper stickers reading "Vote No, Amendment A" on their squad cars, a practice that was later forbidden by headquarters. Spokesmen for the uniformed personnel lashed out against the amendment on the grounds that the pension system's problems were due not to loose eligibility requirements and liberal retirement benefits but to mismanagement by the city government and the pension board. If the amendment was approved, they insisted, the fire and police departments would find it hard to attract qualified officers; and without qualified officers they would have trouble maintaining public safety.[42]

The Cavanagh administration strongly defended the amendment, calling it vital to the city's "future fiscal integrity." The amendment, its supporters claimed, would cut the costs of the firemen's and policemen's pension system, which came to nearly 50 percent of payroll, or more than twice as much as the general employees' retirement plan, which had twice as many active members. By offering veteran officers an incentive to stay on the force, it would also improve the quality of policing and fire fighting. Describing the amendment as "probably the most important money issue Detroit taxpayers have ever been asked to vote on," City Controller Bernard W. Klein warned that if it failed the city would be forced to raise taxes or cut services. The campaign grew acrimonious, with the unions denouncing the city for lying and the city filing charges against the unions with the state's Fair Campaign Practices Commission. As the election approached, Detroit's major newspapers and leading civic groups endorsed the amendment. In a last-ditch effort, roughly 6,000 off-duty firemen and policemen handed out fliers spelling out their case to a quarter of a million households. It was to no avail, however. With heavy support from the city's blacks—who, with Hubert H. Humphrey running for president, turned out at the polls in large numbers—the amendment carried by a narrow majority.[43]

The DPOA made yet another effort to head off pension reform. Shortly after the election, it filed a complaint with the Michigan Employment Relations Commission (MERC) charging that the city had failed to bargain in good faith over the proposed changes in the pension system. The union argued that under the state's Public Employee Relations Act (PERA) the city was obliged to bargain over pensions and that meaningful bargaining could not go on as long as the voters could veto any agreement. The city responded that under the Michigan Home Rule Cities Act, which was enacted long before the PERA, the city had no obligation to bargain over pensions or any other matter covered in the city charter. The legal battle dragged on for five and a half years. The complaint was heard by a MERC trial examiner and, at the request of the union, by the entire commission, whose ruling the city took to the Michigan Court of Appeals and then to the state Supreme Court. The high court, which handed down its ruling in February 1974, agreed with the DPOA that pensions were within the scope of bargaining as defined by the PERA. But holding that a clear statement on the scope of bargaining was not available in 1968, it ruled that the city did not have to renegotiate the changes with the firemen and policemen.[44] The reform measure, which covered all uniformed personnel hired after January 1, 1969, was left intact, at least for the time being.

If the political struggles in Los Angeles, New York, Washington, and Detroit are indicative, the principal constraint on pension reform was the opposition of the uniformed forces. That the firemen and policemen opposed reform is not surprising. After all, it was their retirement income, and thus their own and their families' future well-being, that was at issue in the battles over eligibility requirements, benefit levels, and cost-of-living adjustments. At stake too was their occupational prestige. Generous pensions have long set uniformed officers apart from other municipal employees; they are symbols of the officers' status as the elite of the civil service. What is surprising is that the firemen and policemen were so successful.

Their success was largely a function of their political power, which was formidable. Firemen and policemen probably have more clout than "any other group in city government," C. Erwin Piper said one year after they prevailed on the voters of Los Angeles to lift the ceiling on the cost-of-living escalator. Pension reform is doomed without the support of the uniformed forces, Raymond P. Witte, head of a special committee to revise the New Orleans firemen's and policemen's pension plans, conceded in the late 1960s. And after suffering a setback in a campaign against the "Heart Bill" in the late 1970s, New York City Mayor Ed Koch complained, "You can't get anything through the state legislature in an election year that the unions don't want."[45]

The policemen and firemen derived their political power from several sources. To begin with, they were well organized. Most policemen belonged to independent benevolent or fraternal associations; and virtually all firemen were members of locals of the International Association of Fire Fighters. These groups were long the most cohesive of all municipal employee outfits. They grew even more tightly knit after 1960, when most of them were transformed from pressure groups into labor unions, exclusive bargaining agents for the uniformed forces.[46] Moreover, the firemen's and policemen's unions had a lot of money. It came mainly from members' dues, which were generally deducted from their paychecks and turned over to the unions. To head off pension reform, or to protect other vital interests, the unions raised additional funds—running into the hundreds of thousands of dollars—by levying extra charges on the membership. Finally, firemen and policemen could normally count on the support of other influential groups. Public sector unions usually helped out, not the least because they often cited the firemen's and policemen's pension plans to justify their demands for improved retirement benefits. Private sector unions sometimes lent support too, as did small-town firemen and policemen, who had considerable clout with upstate legislators.

Drawing on these resources, firemen and policemen took an active part in local politics. Their unions lobbied hard in state capitols and city halls. Mayor George Johnston of Tacoma was not exaggerating when he told a reporter in the late 1970s, "Police officers and fire fighters are the biggest lobbyists in the state." The firemen's and policemen's unions also made substantial campaign contributions. Along with other public employee unions, they were among the principal backers of political hopefuls in Boston. And in New York, as Koch remarked in the late 1970s, "Lots of candidates, and certainly a statewide party, cannot run without [their funds]."[47]

Helpful to their friends, the firemen and policemen were rough on their enemies. The case of Ned Shera is noteworthy. A Tacoma Republican, who was elected to the House in 1968 and later assigned to the state Pension Commission, Shera offended the unions by insisting that the legislature should not hike retirement benefits without first estimating long-term costs. In what one state official called "an exercise in political retribution," the firemen and other public employees helped thwart Shera's bid for reelection in 1972. Said one lobbyist, "We did every damn thing we could to bring about his defeat."[48] However rare, such cases reminded the politicians that the unions' opposition might well cost them their jobs.

The chambers of commerce, municipal leagues, and taxpayers associations had a good deal of clout as well. But in the battles over pension reform they labored under a couple of serious handicaps that put them at a disadvantage vis-à-vis the firemen and policemen. One handicap was that these groups usually had a large agenda, which, in addition to reforming the pension systems, generally included improving public services, holding down property taxes, encouraging economic development, and stamping out corruption. These groups could not push elected officials too hard on pension reform for fear of losing their support on other equally important issues. Nor could they come out (or even threaten to come out) against an elected official exclusively on the basis of his position on pension reform. Firemen and policemen had a much smaller agenda, consisting mainly of salaries, benefits, and work rules. Moreover, these matters were often dealt with by different bodies. Thus Los Angeles firemen and policemen could push the City Council very hard on pensions because it had little to say about salaries, which were pegged to prevailing wages in the private sector. Seattle officers could focus their lobbying efforts in the state legislature on pensions because it had little to say about work rules, which were normally hammered out at the bargaining table.[49] With pensions often their highest priority, the firemen and policemen enjoyed a sharp edge over the reformers.

Another handicap was that the commercial, civic, and taxpayers groups had long stood in the vanguard of the nationwide effort to improve policing and fire fighting. Since the turn of the century, they had subscribed to the Progressive position that liberal pensions would help the departments attract qualified recruits, retain veteran officers, and get rid of superannuated personnel. They also held the view that the fire and police departments, if properly staffed and well organized, could make substantial progress in the so-called wars against crime and conflagration.[50] Hence these groups were in a poor position to challenge the uniformed officers when their spokesmen criticized proposals to tighten requirements and lower benefits on the grounds that the changes would make it harder to deal with crime and arson. In point of fact, the firemen and policemen had no evidence that liberal pensions play much of a role in attracting qualified recruits, that younger men make better officers than older men, or even that capable officers have much of an impact on public safety in urban America. But the advocates of pension reform could not raise these issues without repudiating an ideology that they had been wedded to for eighty years.

Besides the clout of firemen and policemen and the handicaps of the commercial and civic groups, there were two other constraints on pension reform, both of which grew out of the peculiar character of public employee pensions. One was that the issue was so complicated that the voters were often at a loss to understand the gravity of the pension problem and to gauge the impact of the proposed changes. Few Americans knew what the actuaries meant by unfunded liabilities; even fewer had a sense of how the actuaries came up with the assumptions on which they based their forecasts. The voters of Los Angeles had no way of figuring out whether Proposition P would increase taxes by more than $5 million per year, as the Property Owners Tax Association of California claimed, or would require no tax hike whatever, as the Fire and Police Protective League assured them. The citizens of Detroit were in an even less enviable position because the campaign over Amendment A was so acrimonious; as City Council President Ed Carey pointed out, "The waters are so muddied by the charges and counter-charges that they don't know what's going on, or which way to vote."[51] With little understanding of the problem and even less of the proposed solutions, the voters tended to base their decisions on short-term concerns about public safety, a tendency that the firemen and policemen exploited very effectively.

The other constraint was that the politicians had little to gain, and possibly much to lose, by taking a stand in favor of pension reform. Consider the options from their perspective. If they rejected the reform proposals, the pension problem would get worse.

A time might even come when the city would have to cut back on essential services in order to pay pension bills. But as pension costs go up gradually, the "Day of Reckoning" would not arrive for a while. By then, another group of politicians would have to deal with the problem. If the politicians went along with the reform proposals, the pension problem might grow less severe. But in most cities, the changes could only be imposed on new appointees. The full impact of the reform would not be felt for a couple of decades, perhaps even longer. Pension reform might ease the fiscal strain, but only for the next generation of politicians. Under these circumstances, many politicians were reluctant to risk antagonizing the policemen and firemen and thus jeopardizing their careers for the sake of pension reform. From their viewpoint, it made more sense to shelve the reform proposals and let another group of politicians grapple with the pension problem or, if the pressure for change grew too strong, as it did in Detroit, to leave the decision up to the voters.

"Nothing is more difficult in a democratic society," political journalist Richard Rovere once wrote, "than to persuade people to deal with a crisis that is not felt but merely anticipated."[52] The firemen's and policemen's pension problem was such a crisis. Many Americans were aware of the problem. But they did not so much feel it—the way they felt inflation, unemployment, high interest rates, and low gas supplies—as anticipate it. The struggle over pension reform bears out Rovere's observation and suggests a few additions. However hard it is to do something about a crisis that "is not felt but merely anticipated" under ordinary circumstances, it is much harder when a few well-organized and well-financed groups do their utmost to head off reform; when reformers are concerned about so many issues that they are reluctant to push elected officials as hard as possible on any one; when citizens cannot fully understand the gravity of the problem, much less gauge the impact of the proposed solutions; and when the risks of dealing with the problem fall on one group of politicians and the benefits accrue to their successors. While these considerations did not make pension reform impossible, they certainly made it very hard.

CHAPTER FIVE

The Legal Constraints on Reform

The political constraints on pension reform were formidable, but they were not insuperable. Despite the strenuous opposition of the uniformed forces, San Francisco voters approved a charter amendment in 1976 that lowered retirement allowances and put a ceiling on the cost-of-living adjustment. A year later, the Washington legislature, deeply troubled by the soaring costs of firemen's and policemen's pensions, overhauled the statewide Law Enforcement Officers and Fire Fighters Retirement System (LEOFF). After much deliberation, the Colorado legislature closed the Denver firemen's and policemen's pension systems to new appointees in 1979 and compelled them to join a much less liberal statewide retirement plan for uniformed personnel.[1]

But if the authorities sometimes overcame the political constraints on pension reform, they rarely got around the legal constraints. They made changes in the pension systems; but as a result of a host of constitutional amendments, state laws, and judicial decisions, the changes applied as a rule only to new appointees. They did not affect current officers, much less retired officers and their survivors, who remained under the provisions of the original plan. So even when reformers are successful, it takes a long time before reforms have much of an impact on the financial condition of big-city firemen's and policemen's pension systems.

The legal constraints on pension reform were an outgrowth of a profound change in the legal status of public employee pensions. Through the early twentieth century, a pension was generally regarded as a gratuity, a reward, or a bounty. "A pension," the Illinois Supreme Court stated in 1905 (and repeated almost verbatim in 1957) "is a bounty springing from the graciousness and appreciation of sovereignty." The provisions of a pension plan, the court ruled nine years later, do not form part of the contract between employee and employer. The District of Columbia Court of Appeals made the same point. "A pension," it declared in 1909, "is not granted because of any property right the pensioner has or may acquire in it, but purely as an act of gratitude from the bounty of the government." It derives solely from "the grace of Congress." It does not matter if the pension fund is contributory, the U.S. Supreme Court held in 1889. His contributions notwithstanding, an employee has no vested right in his pension fund, which is "entirely at the disposal of the government." Until he retires or dies, his interest in the fund is "a mere expectancy." The Minnesota Supreme Court took a similar position, ruling in 1914 that "there is no vested right in the pension accruing in the future from month to month." So did the Texas Supreme Court, which held in 1937 that "as to future installments of a pension the pensioner has no vested right."[2]

Since an employee's rights to a pension are not vested, the courts reasoned, they are not protected by the Fourteenth Amendment, which prohibits the state from taking property without due process of law. Nor, since a pension plan is not part of an employment contract, are these rights protected by Article I (or similar provisions of the state constitutions), which prevents the state from impairing contractual obligations. If an employee has fulfilled the requirements, he is entitled to the payments that have come due; and provided, as the District of Columbia Court of Appeals put it, that the pension law "remains in force and unchanged," he is entitled to future installments. But otherwise, the U.S. Supreme Court ruled, a pension system is "subject to change or revocation at any time, at the will of the legislature." The government, the Illinois Supreme Court held, "has the right to give, withhold, distribute, or recall [a pension] at its discretion." "Because one is placed upon a pension roll under a valid law is no reason why that law may not be repealed and the pension cease." The authorities can not only take away a pension, the Minnesota Supreme Court pointed out, they can even do away with the pension system. They can make these changes at any time, the District of Columbia Court of Appeals declared, "after a party enters the service, either before or after the right to a pension accrues."[3]

Following this reasoning, the courts usually allowed the authorities to change the firemen's and policemen's pension systems as they saw fit. In 1919, they upheld an Illinois law that imposed an age requirement of fifty on Chicago policemen who had hitherto been eligible to retire after twenty years of service at any age. Twenty-five years later, they ruled against a group of Omaha policemen who challenged a charter amendment that raised the age requirement from fifty to fifty-five and that hiked the service requirement from twenty to twenty-five years. In Washington, D.C., the courts gave their approval to a congressional act under which the District ordered a disabled fireman to undergo periodic medical examinations and removed him from the rolls when he refused to do so. In Denver, they found nothing wrong with a state law requiring firemen to report outside earnings and empowering the pension board to reduce the allowance of anyone whose total income exceeded the current salary of the rank at which he retired. A Florida statute that changed the maximum pension for Tampa firemen from three-quarters of final salary to $100 a month—slashing the monthly stipend of one retired officer by more than $200—withstood judicial scrutiny in the 1930s. So did a Texas statute that pegged the retirement benefits of Dallas policemen to the base pay of Dallas patrolmen, a policy that cut the monthly allowance of one retired officer from $183 to $72.[4]

As early as 1917, however, a couple of states began to view public employee pensions in a different way. A pension, the California Supreme Court held that year, is a gratuity only when granted in an ad hoc manner for services previously rendered; under ordinary circumstances it is part of the employee's "contemplated compensation." As the years passed, several other states adopted this view. A pension is "adjusted compensation," ruled the Pennsylvania Supreme Court in 1934; "additional compensation," said the Arkansas Supreme Court in 1939; and "deferred compensation," wrote the Washington Supreme Court in 1956. It is "earned in the present," explained the Pennsylvania court, but "payable in the future," provided the employee meets the eligibility requirements. If a pension is part of an employee's compensation, the courts reasoned, it is part of his employment contract. As the Pennsylvania court remarked, it is "the product of mutual promises between the pensioning authority and the pensioner." The state offers an employee a pension not as an act of charity or benevolence but rather as an inducement to enter the public sector and remain there. Its objective is to enhance the quality of public service. In return for security for himself and his family, the employee agrees to give long and faithful service and to contribute a portion of his salary to the fund. A pension, the courts concluded, is therefore a contractual obligation, which is protected by Article I of the Constitution (and similar provisions of the state constitutions).[5]

At first, only retired employees were covered by this ruling. Before an employee fulfills the eligibility requirements, the Pennsylvania Supreme Court said in 1934, "his retirement pay is but an inchoate right." Once he does so, however, it "becomes a vested right of which the person entitled thereto cannot be deprived." At that time "it has ripened into a full contractual obligation." The New York Court of Appeals agreed. The legislature can change or even abolish a pension plan, it held in 1935. But once an employee has met the necessary conditions, his pension "takes on the attributes of a contract, which, in the absence of statutory reservations, may not legally be diminished or otherwise adversely affected by subsequent legislation."[6]

This ruling was later extended to cover current employees. With the California Supreme Court in the vanguard, several courts took the position that pension rights became vested "at the time the employee enters the public service." The Pennsylvania court ruled in 1954 that the legislature can change a pension system but that the changes "can apply only to conditions in the future, and never to the past." Drawing heavily on several California cases, the Washington Supreme Court held in 1956 that revisions are permissible "only for the purpose of keeping the pension system flexible and maintaining its integrity." A reduction in some benefits must be offset by a corresponding increase in others, it added.[7]

Employing these criteria, the courts struck down one attempt after another to reform the firemen's and policemen's pension systems. In 1936, they nullified a Georgia law that imposed a ceiling on the monthly allowances of retired Atlanta firemen. Fourteen years later they held that the law did not apply to active firemen who were appointed before it went into effect. A Denver charter amendment that abolished the policemen's pension system's cost-of-living escalator fared no better. In 1959, the courts said that the amendment did not affect retired officers; two years later they ruled that it did not apply to active officers who were appointed before it was adopted. It was improper for Scranton to impose an age requirement of fifty-five on current members of the police force, the courts decided. It was also illegal for Long Beach to subject uniformed personnel to a charter amendment that hiked the contribution rate from 2 to 10 percent and replaced a fluctuating pension with a fixed stipend. A 1937 Washington statute that limited the maximum allowance to $125 a month did not affect a Seattle policeman who joined the force in 1925, the courts ruled. And a Tucson policeman who entered the service in 1942 was not subject to a 1952 Arizona law that changed the base on which his allowance was calculated from his last year's salary to the average salary in his last five years.[8]

Some courts have not embraced the view that public pensions are contractual obligations. Until an employee fulfills the necessary conditions, the Texas Supreme Court held in 1970, he has no vested rights in his pension plan, which can be revised and even abrogated by the legislature. Arguing that a pension was neither a gratuity nor a contract, the New Jersey Supreme Court upheld in 1964 a state law that raised the age and service requirements for firemen and policemen. According to the U.S. Court of Appeals, Congress has the power to reduce benefits for both active and retired members of the District of Columbia's firemen's and policemen's pension system. But these courts are the exceptions. The contract theory has prevailed in most states—especially in California, New York, Michigan, Louisiana, Washington, and several others where the big-city firemen's and policemen's pension systems are in serious trouble.[9] And in most of these states it has been a formidable constraint on pension reform.

In several states, the change in the legal status of public employee pensions was the product of a constitutional amendment. New York is a case in point. Not long after the New York City Commission on Pensions came out with its grim report in 1916, the city's firemen and policemen joined forces to head off attempts to revise their pension plans. By dint of vigorous lobbying, they defeated one reform proposal after another in the 1920s and early 1930s. They also prevailed on the state legislature to pass two laws, one in 1924 and the other in 1935, that greatly handicapped the reform effort. The laws provided that the municipal authorities would have to hold a referendum on any change in the firemen's or policemen's pension system if 15 percent of the voters asked for one.[10]

By the late 1930s, however, the uniformed forces were on the defensive. Under strong pressure from the Citizens Budget Commission and other civic groups, many politicians came around to the view that some sort of pension reform was necessary. And in *Roddy v. Valentine*, which was decided in 1935, the New York Court of Appeals let it be known that the courts would not stand in the way. Until an employee meets the eligibility requirements, the high court held, he has no vested rights in his pension fund, which can be modified at any time by the legislature.[11] It was time, New York City's firemen and policemen reasoned, to find another way to safeguard their pensions.

They did not have to look too hard. Under a provision of the state constitution, the legislature was required to submit to the voters in 1936 a proposition calling for a constitutional convention. Overshadowed by the presidential and gubernatorial elections, the proposition did not generate much interest. Nor did it divide the Democrats and Republicans. Governor Herbert H. Lehman and Mayor Fiorello H. LaGuardia favored a convention, while Tammany Hall and the *New York Times*, strange bedfellows, opposed it. The proposition lost heavily upstate, but, with a more than two-to-one majority in the city, it carried by a sizable margin. The delegates were elected in September 1937, and the convention convened in April 1938. The city's firemen and policemen, along with other state and local employees, lost no time in seizing so splendid an opportunity. They quickly drafted an amendment declaring public employee pensions contractual obligations and entitling them to the constitutional protections thereof, an amendment that was sent to the Committee on the Governor and Other State Officers. They then lobbied hard to get it before the convention. And on July 21, the committee submitted to the delegates a constitutional amendment that read as follows: "Membership in any pension or retirement system of the state or of a civil division thereof shall be a contractual relationship the benefits of which shall not be diminished or impaired."[12]

The amendment provoked one of the most heated and protracted debates of the convention. Harold C. Riegelman, a New York City Republican, counsel to the Citizens Budget Commission, and principal spokesman for the opposition, charged that it was "a very ill-disguised raid upon the credit of the State and its municipalities." It saddled them with the unfunded liabilities of the many pay-as-you-go pension systems, a burden of $500 million to $1 billion. Far worse, Riegelman declared, the amendment "absolutely blocks pension reform throughout the State." It not only freezes the eligibility requirements and retirement benefits, but, by placing the "full faith and credit" of the government behind the existing plans, it destroys any incentive to put these plans on a reserve basis. The amendment, Riegelman objected, also encourages the employee groups "to seek constant increases in benefits because every new benefit, no matter how imprudent, becomes contractual and cannot be recalled." To put it simply, it sets up a "one-way street." Under the amendment, "you can build up these benefits as high as you like," but "you can never reduce them." Elected officials can make mistakes, but they can never correct them. Strongly backed by Philip Halpern, a Buffalo Republican and University of Buffalo law professor, Riegelman pleaded with the delegates to turn down the amendment.[13]

Attorney General John J. Bennett, Jr., a New York City Democrat and the amendment's sponsor and chief advocate, responded that the government already had a "moral obligation" to the hundreds of thousands of state and local employees for whom an adequate pension had been a strong inducement to enter the public sector. Most of these employees had assumed that they had vested rights in their pension systems; now that the courts had ruled to the contrary, it was up to the convention to ensure that their rights were protected. This position was endorsed by George R. Fearon, a Syracuse Republican and one-time Senate majority leader. Bennett also challenged Riegelman's assertions that the proposed amendment would increase the taxpayers' burden. Once the cities realize that they have to make good on their promises, they will stop the deplorable practice of "piling up" benefits without building up reserves. They will have no choice but to put their pension systems on a funded basis. Moreover, the attorney general reminded the delegates, "this amendment protects only those who are presently in service." Nothing in it requires the state or its municipalities to provide retirement benefits to new employees, much less to improve these benefits. Strongly supported by Fearon, Bennett called on the convention to adopt the amendment.[14]

The issue was joined on the day the amendment was placed before the delegates. After spelling out his objections, Riegelman offered what was in effect a substitute amendment. It struck out the reference to contractual relationships and provided merely that in a funded system the reserves "shall be maintained without impairment." That, said Riegelman, "is as far as we should go." Bennett opposed the amendment, as did State Senator Benjamin F. Feinberg, spokesman for the Committee on the Governor and Other State Officers. The delegates, who were under great pressure from the policemen, firemen, and other public employees to back the original amendment, turned down Riegelman's revision.[15]

The issue came up again a few weeks later. In an effort to limit the impact of Bennett's proposal, and perhaps to win additional votes for the opposition, Riegelman offered a new amendment. This one stated that public pensions were contractual relationships, but only if the retirement system operated on a reserve basis. The amendment would have covered most state and municipal systems for nonuniformed employees. But it would not have applied to most local firemen's and policemen's funds, including the New York City firemen's and policemen's funds, which were still on a pay-as-you-go basis. A clever gambit to divide the public employees, Riegelman's amendment nonetheless went down to defeat by a margin of 53 to 43.[16]

Shortly after, widespread opposition to the original amendment surfaced outside the convention. Henry J. Amy, executive director of the Citizens Budget Commission, warned that the amendment would block pension reform in the state, put an intolerable burden on the taxpayers, and perhaps even drive the municipalities "to default on their obligations." Citing the enormous liabilities of New York City's firemen's and policemen's pension systems, the *Times* called the amendment "a Dangerous Proposal," a "sop" to the uniformed forces, which "ought to be killed." The *New York Sun*, while conceding that public employee pensions should be protected, contended that the attorney general's amendment "is not the way to do so." Perhaps encouraged by this outpouring of opposition, Riegelman made a last-ditch attempt to head off the amendment. This time, he pushed for a proposal that not only defined membership in a reserve system as a contractual relationship but also treated employee contributions in a pay-as-you-go system as a "trust obligation." Bennett spoke against it, as did former Governor Al Smith. With the majority holding firm, the delegates rejected the proposal by a margin of more than two to one and adopted the original amendment by a vote of 112 to 43.[17]

Although the city's civic and commercial groups were sharply opposed to the amendment, they were hard put to mount an effective campaign against it. For the convention had lumped the proposal together with forty-nine other largely unrelated and allegedly noncontroversial proposals in an omnibus amendment called Amendment No. 1. As a package, the amendment divided the civic and commercial groups. The Brooklyn Real Estate Board and the City Club of New York came out against it, citing as their main objection the proposal putting pensions on a contractual basis. The Citizens Budget Commission endorsed it on the grounds that the disadvantages of Bennett's proposal were outweighed by the advantages of the proposals providing for tax and fiscal reform. The New York City Real Estate Board came to a similar conclusion. Thus Amendment No. 1 turned out to be far from noncontroversial. In favor were the Democratic and Republican parties, the *Herald-Tribune* and *Daily News*, Al Smith and Thomas E. Dewey. Opposed were Fiorello H. LaGuardia, the *Times* and *Sun*, the American Labor party, and the American Civil Liberties Union.[18] In the end, the voters adopted Amendment No. 1 by a 54 percent majority—a vote that gives no insight into what they thought about the change in the legal status of public pensions.

A handful of states followed New York's lead, although not immediately. Michigan adopted a constitutional amendment, drafted by the convention of 1962, that defined pension benefits as a contractual obligation, "which shall not be diminished or impaired." The amendment, which generated little debate, also provided that henceforth all state and local retirement systems had to operate on a funded basis. Illinois gave its approval to a new constitution in 1970, one provision of which was an almost verbatim copy of New York's constitutional amendment. Passed by the convention after a brief debate, the provision was especially noteworthy because the Illinois Supreme Court had long held the traditional view that mandatory participation in a pension system conferred no vested rights. Louisiana ratified two constitutional amendments in 1973, one for teachers and the other for other public employees, that stated that public pensions were contractual obligations. Drafted by the 1973 convention, the amendments even provided that the state would guarantee payments due an employee on his retirement (or his dependents on his death). Alaska and Hawaii amended their constitutions in the same way, although in New Jersey a proposed amendment, modeled closely on New York's and strongly supported by firemen, teachers, and other public servants, died in committee at the Constitutional Convention of 1947.[19]

In a few states, the legal constraints on pension reform were the result of a legislative act. Wisconsin is a case in point. In January 1945, State Senator Bernhard Gettelman announced that he would introduce a bill imposing a ceiling of $125 to $150 a month on the retirement allowances of Milwaukee's nonuniformed employees. In defense of the measure, he said that the pension funds have become "a Frankenstein monster that might very well destroy the taxpayers." Shortly after, Assemblyman Charles F. Westfahl, expressing concern for the long-term solvency of the city's firemen's and policemen's pension systems, served notice that he would file legislation limiting monthly benefits for uniformed officers to $150. (Westfahl had submitted a similar bill two years earlier, but without success.) As far as Milwaukee's public employees were concerned, these proposals could not have come at a worse time. Two years before, the Milwaukee Government Service League, which consisted of 7,500 city and county employees, underwrote a suit challenging a state law that lowered death benefits for Milwaukee employees. While the trial court ruled in the league's favor, the Wisconsin Supreme Court, standing by previous decisions, reversed the ruling. Pointing out that public pensions were not contractual relationships, the court held that the legislature had the right to change the benefits as it saw fit, an opinion that was subsequently upheld by the U.S. Supreme Court.[20]

Spokesmen for the public employees lashed out against the proposed ceilings. Speaking for the Milwaukee Government Service League, City Librarian Richard E. Krug labeled them "breaches of faith" that would weaken the public service. Chief of Police Joseph T. Kluchesky warned that they would wreck the force by driving most ranking officers, himself included, to retire before the ceilings were imposed. In the meantime, the employee groups decided that the most promising way to safeguard their interests was to ask the legislature to enact a law stating that public pensions are contractual obligations. These groups had reason to believe that, if the legislature went along with this request, the courts would not permit succeeding legislatures to lower benefits for current employees. The league, the Wisconsin Association of Police Chiefs, the Milwaukee Board of Supervisors, and the American Federation of State, County and Municipal Employees all drafted bills. Their efforts were strongly endorsed by the Wisconsin State Federation of Labor and other private sector unions as well as by the *Milwaukee Sentinel*, which attacked the authorities for breaking promises to their employees. The campaign built up so much momentum that in February 1945 all seven Milwaukee state senators—including Gettelman—jointly sponsored a bill safeguarding the pension rights of city and county employees. Introduced on behalf of the Government Service League, the bill was referred to the Senate Committee on State and Local Government.[21]

Although the county supervisors were in favor of the bill, the City Council had strong reservations. The council was worried that the measure might exacerbate the financial problems of the firemen's and policemen's pension funds. Supposedly operating on a reserve basis, the two systems had run up a deficit of about $15 million, largely because the council had not made the recommended contributions. The council also was afraid that the bill might prevent reforms then under consideration. Among them was a proposal to base the officers' retirement allowances on their last five years' salaries instead of their salaries at retirement, a proposal designed to deter widespread retirement when a wage hike went into effect. A few councilmen were concerned as well that the bill might jeopardize the attempt to create a single citywide retirement plan for all new employees, uniformed as well as nonuniformed. This scheme—which had been tried in Boston, Cincinnati, and San Francisco—would presumably lower pension costs. These reservations were so strong that in late March the council refused to endorse the bill. Instead, it decided to ask the legislature to refer the proposal to an interim committee that had been established to look into the feasibility of a statewide retirement plan for all public employees, a position backed by the League of Wisconsin Municipalities but criticized by the employee groups and the *Milwaukee Sentinel*.[22]

The Committee on State and Local Government approved the bill on March 21, but, deferring to the wishes of the council, the Senate put it off for two weeks. The public employees brought enormous pressure on the council to reverse its position. But after receiving a grim report from the actuary about the financial condition of the firemen's and policemen's pension systems, the council stood firm. It also appointed a special committee to study the city's pension plans. After consulting with George B. Buck, the committee offered the firemen and policemen a deal. If they would back a proposal to require all new appointees to join the city employees' retirement plan, the committee would support the bill to put the pension systems on a contractual basis. But the uniformed forces turned the offer down, and the council later repudiated it. In the meantime, the public employees exerted intense pressure on the legislature to move ahead on the bill. With the support of the county supervisors, a law covering county employees was enacted. But urged by the City Council to wait until its special committee finished its study, the legislature delayed action on a bill dealing with city employees. On May 10, by which time it was apparent that the city's leaders were unable to reach a consensus and that several councilmen were strongly opposed to the bill, the legislature turned the matter over to its interim committee.[23]

From the outset, the interim committee was caught up in the controversy over the legal status of public pensions. On one side were the public employees, who urged the committee to endorse the pending legislation. The Milwaukee police "have a promise,"

Chief Kluchesky testified, "now they want a contract." On the other side were the civic groups, which strongly opposed the bill. An actuarial study should be made first, said a spokesman for the City Club, who also argued that the fire, police, and general employees' plans should be merged into one system. Late in May, the employee groups prevailed upon the Senate to remove the bill from the committee; only swift action by the City Council prevented it from coming to a vote. But while the bill was no longer on its agenda, the committee was still pressed to take a stand on it. Later it was also dragged into a controversy over whether the state should delegate authority over Milwaukee's pension plans to the city, a change favored by the City Council and opposed by the employee groups. Despite these distractions, the committee got on with its business. After a year of work, it came to the conclusion that the local pension systems left much to be desired; it recommended that the legislature get rid of them and in their place set up a statewide retirement plan that would cover all state, city, and county employees—except teachers. Early in 1947, the committee drafted a bill that incorporated these recommendations.[24]

The bill helped forge a consensus in Milwaukee, although not one in support of the committee's position. Faced with the chance of being moved into a statewide plan, the firemen and policemen softened their opposition to a citywide system. As the prospects of a consolidated system rose, the civic groups toned down their objection to putting pensions on a contractual basis. Opposed to state takeover of the city's pension systems, the council was anxious to find an alternative to the committee's proposal. With the help of a citizens committee, the council came up with a plan of its own in March 1947. The city's firemen's and policemen's pension systems would be closed to new entrants, who would be covered by the city's general employees' retirement plan; this plan would give them higher benefits than nonuniformed employees but lower benefits than current firemen and policemen. Whichever plan an employee belonged to, his pension would henceforth be a contractual obligation. This plan was incorporated into three bills that were introduced into the legislature as amendments to a bill exempting Milwaukee employees from the proposed statewide retirement system—a system that was approved by the Senate in late April. Testifying before the Joint Committee on Finance in early May, Milwaukee officials stressed that this plan would enable the city not only to reduce the deficits of the firemen's and policemen's pension plans but also to put its entire retirement system on a sound fiscal basis.[25]

Most elected officials, public employees, and civic groups endorsed the council's proposal, although the Milwaukee Citizens' Governmental Research Bureau objected on the grounds that it would place a staggering burden on the taxpayers. Impressed by such widespread agreement on what was usually so divisive an issue, the legislature excluded Milwaukee's public employees from the statewide retirement plan, which was adopted by the Assembly in May and approved by the governor in June. After a heated debate over an amendment to include Milwaukee's elected officials in the citywide pension plan—an amendment that eventually passed—the legislators also gave their approval to the City Council's bills. At the same time, the Senate turned down another amendment to impose a ceiling of $150 a month on retirement allowances for city employees, a revised version of the proposal that sparked the controversy over the legal status of public pensions. With representatives of the city government, civic groups, and employee associations looking on, Governor Oscar Rennebohm signed the legislation on July 24, 1947, two and a half years after the campaign for contractual status got under way.[26] The city's firemen, policemen, and nonuniformed employees now had vested rights in their pension plans, rights enjoyed by Wisconsin teachers and Milwaukee County employees but not by most of the state's other public servants.

Ohio and Massachusetts took much the same tack as Wisconsin. In Ohio, the state legislature passed two statutes in 1937 that gave firemen and policemen vested rights in their pension plans on retirement and barred any reduction in benefits afterward. The statutes were designed to circumvent the Ohio Supreme Court's recent rulings that a pension was a gratuitous allowance that could be reduced or even terminated by the government either before or after retirement. After the statutes went into effect, the court held that the legislature could still lower benefits, tighten requirements, or otherwise revise the plans—but only for active members. In Massachusetts, the state legislature enacted a law in 1956 that defined pensions as a contractual relationship and applied the definition to active as well as retired employees. The statute was recommended by a special legislative commission that was unhappy with the Massachusetts judiciary's view that public employees had no vested rights in their pension plans. The statute was so restrictive that in 1973 the Supreme Judicial Court ruled that a bill to hike employee contributions without raising retirement benefits was invalid except as applied to prospective employees.[27]

In most states, the legal constraints on pension reform were the result of judicial decisions. California, Georgia, and Pennsylvania led the way, followed by Washington, Colorado, Idaho, Montana, Arizona, Arkansas, and Alabama. In these states, the court abandoned the traditional view—that an employee's rights in his pension system are not vested—and adopted what is now the prevailing view—that not only are an employee's rights vested but they become vested when he enters the service.[28]

The traditional view rested largely on the assumption that pensions are gratuities. They spring "from the graciousness and appreciation of sovereignty," the Illinois Supreme Court said in 1905; they are given "purely as an act of gratitude from the bounty of the government," noted the District of Columbia Court of Appeals four years later. Developed in the late nineteenth century, long before retirement systems became a common feature of American life, this assumption had its origins in Europe, where royalty had long provided pensions to retainers for faithful service. It probably took hold in the United States because the courts were reluctant to define property rights in so broad a way as to upset traditional notions of sovereignty and impose severe restrictions on government. Yet if the assumption's roots are murky, its consequences are clear. If pensions are gratuities, the courts reasoned, it follows that the government can change or even abrogate a pension system at any time without the consent of the employees.

From the outset, the traditional view was attacked on the grounds that employees acquired vested rights in their pensions by virtue of having made regular contributions to the funds. By reducing their benefits, the government deprived them of property without due process of law, a violation of the Fourteenth Amendment. The issue came before the U.S. Supreme Court in 1889. The case, *Pennie v. Reis*, was brought by the estate of a San Francisco policeman, who died shortly after the state legislature dropped from his pension plan a death benefit of $1,000. The plaintiff contended that the deceased had a vested right to the $1,000 because he had put $2 a month into a fund since 1878. In a landmark decision, the court rejected this position. Pointing out that the $2 was deducted from the officer's salary, it held that the money did not belong to him. "Though called part of the officer's compensation, he never received it or controlled it. . . . He had no such power of disposition over it as always accompanies ownership of property." Had he been paid the $2 and then contributed it on condition that a death benefit would be given to his survivors, "a different question would have been raised." But in the case at hand, the money belonged to the state, the fund "was entirely at [its] disposal," and by changing the law, the legislature "impaired no absolute right of property."[29]

The traditional view was also attacked on the ground that the employees had made a contract with their employer upon entering service, of which the pension plan was an integral part. By revising the plan without their consent, the government impaired a contractual obligation, thereby violating Article I of the Constitution. The courts rejected this argument as well. There is no contractual relationship between a city and its employees, the Wisconsin Supreme Court held in 1904. An office is not the property of the officeholder; it is a public trust in which he has no vested rights. An employee's pension, like his wage or tenure, is a matter of legislative policy. Public officers, the Illinois Supreme Court declared in 1914, "are wholly within the control of the Legislature, which may at pleasure create or abolish them, modify their duties, shorten or lengthen their terms, increase or decrease the salary, or change the mode of compensation." The U.S. Supreme Court adopted the same position in 1937. The New Jersey Supreme Court ruled in 1904 that a pension is a contractual relationship, but only where the plan is voluntary and the allowances are not paid out of public funds.[30] Since few public employee pension systems fell into these categories, the authorities had the right in most cases to make whatever changes they saw fit.

Although the traditional view withstood these attacks, it eventually succumbed to other challenges. One reason for its demise is that the courts gradually found the gratuity theory untenable. In the first place, some states had constitutional provisions that barred the legislature from giving public funds to private parties. Typical of these provisions was a clause of the Georgia constitution that read, "The General Assembly shall not, by vote, resolution or order, grant any donation or gratuity in favor of a person, corporation or association." These restrictions, most of which were imposed in the second half of the nineteenth century, were designed to prevent elected officials from making extravagant gifts to railroad corporations, a practice that left many states with large long-term debts. If a pension was merely a gratuity, the California Supreme Court pointed out as early as 1917, it would be unconstitutional, a position the court was reluctant to embrace. The courts made the same point in Georgia (1950), in Washington (1956), and in Arizona (1965)—where they ruled that "pensions cannot be sustained as constitutional unless anchored to a firmer basis than that of a gift."[31]

In the second place, the gratuity theory had a decidedly regal cast to it. As the world's foremost republic, the United States was a singularly inappropriate setting for what the Pennsylvania Supreme Court referred to as "the ancient idea that a pension is a manifestation of sovereign generosity"—a concept that "has come down through the centuries wearing a cloak of monarchical dispensation." The evocation of what Rubin G. Cohn called the "archaic imagery of a compassionate and generous king rewarding his humble and devoted subjects" had an odd ring to it in a country that was not ruled by a king and whose residents were citizens, not subjects, much less "humble and devoted subjects." As the years went by, the gratuity theory grew more than a little anachronistic. The concept of sovereignty, on which the theory rested, lost much of its force in the early twentieth century, when the authorities adopted civil service, standardized salaries, and other reforms that reduced their autonomy. What force the concept retained was further weakened a generation later, when the authorities recognized public sector unions, set up formal grievance procedures, and bargained collectively over salaries, benefits, and working conditions. By the second half of the twentieth century, it seemed highly anachronistic for the courts to speak of a pension as "a bounty springing from the graciousness and appreciation of sovereignty."[32]

In the face of these constitutional and ideological problems, a good many states disavowed the gratuity theory. A pension is a gratuity, the California Supreme Court ruled in *O'Dea v. Cook* in 1917, "only where it is granted for services previously rendered, and which at the time they were rendered gave rise to no legal obligation." Where the services are rendered under statute, "the pension provisions become a part of the contemplated compensation for those services." The Supreme Court of Pennsylvania took much the same view in 1934. "Where an allowance is made out of hand, gratuitously, and purely for past services," it held in *Retirement Board of Allegheny County v. McGovern*, it is a pension. Otherwise it is "retirement pay." And retirement pay, the court pointed out, is " 'adjusted compensation' presently earned, which, with contributions from employees, is payable in the future." The courts arrived at much the same opinion in other states. A pension was defined as "increased compensation" in Minnesota in 1928, "additional compensation" in Arkansas in 1939, and "deferred compensation" in Washington in 1956.[33] Once the courts decided that a pension was part of an employee's compensation, they were well along the way to the conclusion that he had vested rights in his pension plan.

Another reason for the demise of the traditional view was that the Supreme Court's reasoning in *Pennie v. Reis* left much to be desired. "What one finds in this landmark decision," Cohn observed, "is a judicial confrontation with a novel conflict and its resolution by a strained application of conventional legal concepts." "Even in 1889," he pointed out, "principles of ownership of property, intangible as well as tangible, did not demand, as an indispensable condition, that the owner have physical custody, control, or possession of the res."[34] As the years went by, the high court's reasoning grew even more strained. By the second third of the twentieth century, it was routine for employers to make deductions from employees' paychecks for Social Security, income taxes, and insurance premiums. While the employees never possessed the money, there was no doubt that it belonged to them. It was also common for affluent Americans to set up trusts for their children and grandchildren as a means of avoiding estate taxes. Although the beneficiaries often had no say about the disposition of the trust, there was no denying that it was their property.

Pennsylvania and Georgia were among the first states to repudiate the principles of *Pennie v. Reis*. In *Retirement Board of Allegheny County v. McGovern*, the Pennsylvania Supreme Court held in 1934 that an employee's contributions "are as much wages or salary when deducted at the source as though they had been paid directly." To deduct a portion of an employee's paycheck for a retirement fund is to take part of his salary and not to reduce his salary. "To take this earned salary and deny any vested right in it is against fair dealing," the Pennsylvania court argued. To hold otherwise would be "unjust," especially in the case of a retired employee who has made contributions for years and now "is beyond an age when his usefulness as a provider is intact." The Georgia Supreme Court took a similar position in *Trotzier v. McElroy* in 1936, a position it spelled out more fully in *Bender v. Anglin* fourteen years later. In *Bender*, a case brought by a retired Atlanta fireman, the court refused to go along with the view that his contributions were not his property. "When it [the money] was taken and placed in the fund from which the benefits were to be paid," the court said, "it constituted a contribution by the fireman of a part of the salary he had earned by his services."[35]

The repudiation of the principles of *Pennie v. Reis* was especially noteworthy in Colorado. Through the 1940s, the Colorado Supreme Court adhered closely to *Pennie v. Reis,* ruling in the *Albright* and *Behrman* cases—both of which involved the Denver firemen's pension fund—that an employee's contributions were not his property and did not give him vested rights in his pension. In 1959, however, the court changed its mind. The case, *Police Pension and Relief Board of City and County of Denver v. McPhail,* was brought by a group of retired Denver policemen. The plaintiffs challenged the validity of a 1956 charter amendment that abolished a cost-of-living escalator that pegged retirement allowances to wage hikes. Siding with the officers, the court overruled *Albright* and *Behrman,* saying that the "faulty premise" in these cases was "the conclusion that deductions from the salary of the member are not his property." *Pennie v. Reis* notwithstanding, possession is not "a proper test" of ownership. Because an employee never holds the sum deducted from his salary does not mean that it "is not his property and does not constitute an actual consideration paid into the fund." Income taxes are paid on the entire salary, the court pointed out. Although compulsory, deductions for pension funds are as much an employee's property as deductions for income taxes. Once this principle is recognized, the court concluded, "the relationship between the City and the member acquires the character of a contract."[36]

Still another reason for the demise of the traditional view is that the courts gradually redefined the nature of public employment. Through the late nineteenth century, the courts adhered to the doctrine that public office is a privilege, not a right. It is a grant by the government, not the property of the officeholder. The terms and conditions of public employment are therefore legislative matters that the authorities can change as they see fit. Justice Oliver Wendell Holmes stated this view succinctly in 1892. Dismissing a plea by a New Bedford policeman who was fired for violating a regulation that restricted his political activity, Holmes argued, "the petitioner may have a constitutional right to talk politics, but he has no constitutional right to be a policeman." In the twentieth century, however, the doctrine of privilege gradually gave way to the concept of substantial interest. According to this concept, a public office is less than a right but more than a privilege. Even in the absence of a right to public employment, the employees have other rights that cannot be abridged by the government without compelling reason. The concept of substantial interest enhanced the civil liberties of public employees, especially freedom of speech, right to privacy, and privilege against self-incrimination.[37] It also helped to remove the principal objection to the notion that a contract of employment exists between the government and its employees.

A pension plan is part of the employment contract, the courts held, not only because a pension is part of an employee's compensation but also because, as the California Supreme Court put it in 1947, it is designed to attract qualified personnel to public service. "It is obvious that this purpose would be thwarted if a public employee could be deprived of pension benefits and the promise of a pension annuity would either become ineffective as an inducement to public employees or it would become merely a snare and a delusion to the unwary." The Supreme Court of Georgia made the same point in 1950. "No amount of judicial theorizing can escape the plain fact that the existence of a law which promises benefits upon retirement or injury constitutes an inducement to accept the position of fireman, thereby incurring the risks and hazards which the very nature of the work imposes." A retirement plan, the Pennsylvania Supreme Court declared in 1954, is one of the cardinal reasons an individual enters public service instead of private enterprise, which as a rule offers higher salaries and better opportunities. In return for long and faithful service, it promises him "economic recognition in the evening of his life when he has given with full will of his energies in the morning and afternoon."[38]

A pension is therefore a contract, the Pennsylvania Supreme Court said in 1954, not "a grant of the Republic" or "a gift of the City Fathers," but rather "the product of mutual promises between the pensioning authority and the pensioner." As such, it is protected by Article I of the Constitution, which prohibits the state from impairing contractual obligations, and cannot be modified without the consent of both parties. For a while, the courts held that an employee's retirement pay was "an inchoate right" until he has satisfied the eligibility requirement, at which time it became "a full contractual obligation." But this notion was inconsistent with the view that the pension plan was an integral part of the employment contract and a principal inducement to join the service in the first place. With the California Supreme Court in the lead, the courts soon came around to the position that an employee's pension rights vest at the time of his appointment. "The promise on which the employee relies is that which is made at the time he enters employment," the Washington Supreme Court held in 1965 in *Bakenhus v. City of Seattle*, "and the obligation of the employer is based upon this promise."[39]

Although an employee has vested rights in his pension, the authorities can still change the plan, the courts pointed out. But as the Pennsylvania Supreme Court held in 1954, the changes "can apply only to conditions in the future, and never to the past"; they can be imposed on new members, but not on current employees, much less on retirees or their survivors. "Whether it be in the field of sports or in the halls of the legislature," the court went on, "it is not consonant with American traditions of fairness and justice to change the ground rules in the middle of the game." The Supreme Court of Arizona, which held that a pension is "a firm, binding contract," arrived at the same conclusion.[40]

The California Supreme Court took a somewhat less stringent line. While holding that a pension is a contractual obligation, it argued in 1947 that an employee is not entitled "to any fixed or definite benefits, but only to a substantial or reasonable pension." The legislature can modify a pension system in order to maintain its integrity and enable it to adjust to changing conditions. But the modifications can only be imposed on current employees if a change that has an adverse effect on the employees is accompanied by a change that has a beneficial impact.[41] The California position was later adopted by the courts in Washington, Colorado, and several other states.

Some courts have refused to go along with what the Washington Supreme Court called "the more enlightened trend." The Nebraska Supreme Court has not overruled *Lickert v. City of Omaha*, a 1944 opinion that held that pensions are not contracts, that pension contributions are not the employee's property, and, following *Pennie v. Reis*, that legislative modifications impair "no absolute right of property." The Supreme Court of Iowa has not overturned *Talbott v. Independent School District of Des Moines*, a 1941 decision that said, "The fact that these retirement or disability payments are not gratuities, is not, in our judgment, sufficient to give to them the character of a property, or a vested right, or a contract right, which cannot be adversely affected by subsequent legislation by either the state or a municipality." The Texas Supreme Court has taken much the same stance. The New Jersey Supreme Court has adopted what some consider a more sophisticated view. Taking exception to the common practice of using labels such as "gratuity" and "contract," it argued in 1964 that pensions are much too complex to be summed up "in one crisp word." Pointing out the inadequacies of both theories, the court concluded that the legislature can modify a pension plan without necessarily violating the constitutional rights of current employees.[42]

But in many courts, "the more enlightened trend" has caught on. And even when these courts have been divided, it has not been over the contract theory but over its application. In *Bakenhus v. City of Seattle*, for example, the Washington Supreme Court struck down a 1937 statute that imposed a ceiling of $125 a month on the retirement allowances of Seattle policemen. The court held that the law impaired the plaintiff's contract with the city on the grounds that it reduced his pension by roughly one-third without providing any corresponding increase. Justice Matthew A. Hill filed a strong dissent, in which Chief Justice Frederick G. Hamley concurred. Hill objected not so much to the majority's holding that a pension was a contract as to its ruling that the plaintiff had not consented to the imposition of the ceiling. As he understood it, the court's position was that if the authorities modify the pension system in a way that is favorable to the employee, "by his silence and continuing employment he assents to the modification." But if the change is (or turns out to be) unfavorable, the employee "can at any time, however remote, assert that he never consented to the change and that it impairs the obligations of his contract." This position, which Hill described as "Heads, I [the employee] win; tails, the city loses," is untenable. The plaintiff's suit should have been rejected because when he accepted benefits under the 1937 act, "he acquiesced in all the modifications [it] made in his contract unless he in some way made known the specific modifications to which he did not assent."[43]

As the state courts came around to the contract theory, the federal courts emerged as the stronghold of the traditional view. In a long line of rulings—from *Pennie v. Reis* (1889) to *Dodge v. Board of Education of Chicago* (1937) to *Flemming v. Nestor* (1960)—the U.S. Supreme Court insisted that an employee has no vested rights in his pension. In 1974, however, the U.S. District Court for Texas, Western District, handed down a decision that suggests that the traditional view may be on the wane in the federal courts. The case, *Muzquiz v. City of San Antonio*, was brought by a group of former San Antonio firemen and policemen who left the service without a pension and argued that a state law that denied them a refund of their contributions deprived them of property without due process. Relying on *Pennie v. Reis*, the court ruled against the officers on the grounds that they could not recover that which did not belong to them in the first place. But it also called the *Pennie* rule "harsh," adding that it was constrained "to apply the law as it is given to us, *stare decisis*." "If this were a case of first impression, this Court might possibly have arrived at a different result." But it was not. Thus the court was compelled to abide by eighty-five years of constitutional law at the federal level.[44] If *Muzquiz* is indicative, the federal courts may eventually abandon the traditional view. But in the meantime they are bound by a ruling that has long been rejected by most state courts.

The contract theory has not been applied uniformly. In New York, the courts interpreted it quite strictly. They even held that the trustees of a pension fund cannot employ an updated mortality table to compute retirement allowances for current members if the result is to reduce their monthly payments. In Michigan, the Supreme Court ruled that a similar constitutional amendment does not prevent the legislature from hiking the employees' contribution rate. And yet in Massachusetts, where employees' rights are vested under state law, the Supreme Judicial Court said that a bill to raise the contribution rate impairs the state's contractual obligation and deprives the employees of property without due process. The Arizona Supreme Court ruled that the legislature can make no changes without the employees' consent. But the Pennsylvania Supreme Court backed away from so restrictive a position. It took the view that while the authorities cannot reduce pension benefits, they can tighten eligibility requirements and raise contribution rates, provided that the revisions are meant to enhance the actuarial soundness of the retirement system.[45]

Although it has not been applied uniformly, the contract theory has nonetheless operated as a strong constraint on pension reform. The courts have cited it as grounds for striking down efforts to impose a ceiling on retirement allowances in Atlanta, to abolish a cost-of-living escalator in Denver, to hike contribution rates in Long Beach, to impose an age requirement in Scranton, and to redefine final salary in Tucson. Moreover, as the courts have made it clear that under the contract theory the authorities can make few if any changes in the status of current employees, reformers have been forced to focus their efforts exclusively on new appointees.

While reformers have brought about noteworthy changes in Detroit, Oakland, Minneapolis, and other cities, it will be a long time before these changes have much of an impact on the fiscal condition of the firemen's and policemen's pension funds. Retired officers will continue to receive their regular benefits. And current officers, many of whom will not retire for ten or twenty years, will be entitled to the same benefits, which they and their survivors will collect for another twenty or thirty years. Thus, while the contract theory has not absolutely blocked pension reform, as Harold Riegelman claimed at the New York State Constitutional Convention of 1938, it has made it much more difficult to bring about.

The contract theory has also made it much more difficult to preserve reform. As Riegelman rightly pointed out, it sets up a "one-way street." So far as current employees are concerned, the government can raise retirement benefits but cannot lower them, can loosen eligibility requirements but cannot tighten them. The contract theory does not stop the authorities from imposing more stringent conditions on new appointees; but if they abolish or water down these conditions, they cannot later reimpose them, even on employees who entered the service under the more stringent conditions. Once the authorities rescind a reform measure, their action is just about irreversible.

There is something perplexing about the contract theory. As the Supreme Court of Pennsylvania said in 1934, it may be that persons who have grown old in the public service "should be protected and cared for by the governmental agency in whose behalf they have so long labored." It may also be that the promise of a liberal pension is one of the principal inducements to enter public service and to remain in it. And as the Pennsylvania Supreme Court argued in 1954, it may be that "it is not consonant with American traditions of fairness and justice to change the ground rules in the middle of the game."[46] If so, it is clear that an employee is entitled to the benefits that were promised him when he entered the service. But it is not clear that he is entitled to the improved benefits granted afterward. It is hard to see why a New York City policeman who was appointed in 1950, when retirement allowances were based on his last five years' salary, should have vested rights in the 1963 amendment that based allowances on his final salary. It is also hard to see why a Los Angeles fireman who was hired in 1955, when the pension provided no cost-of-living adjustment, should have vested rights in the 2 percent escalator adopted in 1966 and the full escalator approved five years later. By adopting the contract theory, the courts have made it easier to protect the legitimate expectations of the employees; but by defining these expectations so broadly, they have made it harder to maintain the financial solvency of the pension systems.

The contract theory, it might be argued, should tend to undermine the political constraints on pension reform. If the reforms can only be imposed on new appointees, current members do not have that much at stake; at any rate, they have less at stake than if the reforms could be imposed on them as well. If so, they should be reluctant to go to great lengths to stymie reform. The Cavanagh administration acted in 1968 on that assumption. In an attempt to overcome opposition to Amendment A, which tightened the eligibility requirements and capped the cost-of-living escalator, the mayor sent a letter to firemen and policemen saying that even if the amendment carried, they had the option of remaining in the old pension system.[47] The argument does not hold up, however. Detroit's firemen and policemen strongly opposed the Cavanagh administration's reform proposal at every turn—first in the council, then at the polls, later before the Michigan Employment Relations Commission, and finally in the courts. It is possible that uniformed officers might have resisted even more strongly if the authorities had attempted to impose the reforms on them as well, but it is hard to imagine what else they could have done.

Current officers have strongly opposed attempts to tighten eligibility requirements and/or reduce retirement allowances for new appointees for two main reasons. One is financial. If reform proposals are adopted, recruits will soon find out that veteran officers, who work no harder than they do, have a more liberal pension plan. The discovery will probably prompt them to mount a campaign for parity, a campaign that will inevitably lead them to place a higher priority on improving pensions than on raising salaries, the principal objective of their fellow officers. Thus current officers fear that pension reform will divide the uniformed services in a way that will make it harder to win wage hikes, thereby holding down salaries (and, in the long run, retirement allowances). The other reason is institutional. Recruits will no doubt bring enormous pressure on rank-and-file organizations to support the campaign for parity; they may even form insurgent groups along the line of Pension Forum. Current officers are well aware that this pressure may divide their organizations, undermine their leadership, and weaken their position vis-à-vis city and state officials. They fear that pension reform might produce the sort of internecine conflict that nearly destroyed the New York City Patrolmen's Benevolent Association in the 1940s.

Far from undermining the political constraints on pension reform, the contract theory tends to reinforce them. In the mid-1960s, for example, the Washington Public Pension Commission attempted to consolidate the local firemen's and policemen's pension plans into a statewide retirement system. As the Washington Supreme Court had recently adopted the contract theory, the commission made no effort to force current employees into the new system. Even so, the big-city firemen and policemen opposed the change; since their own pension plans could not be modified without their consent, they had no incentive to compromise. In the end, they went along with consolidation, but only after the commission proposed a new system that was even more liberal than the old one.

An instructive contrast is provided by the successful campaign to replace the local firemen's and policemen's pension plan with a statewide retirement system in Minnesota in the late 1970s. There, too, the big-city firemen and policemen opposed the change, which was proposed by the Legislative Commission on Pensions and Retirement. But since the Minnesota courts had not adopted the contract theory, the officers realized that if they remained intransigent, the reformers might try to reduce their pensions. Ultimately, they went along with a proposal that left their own plans unchanged but forced new appointees into a statewide system that tightened the eligibility requirements and abolished the automatic cost-of-living adjustment.[48]

In addition to the political constraints on pension reform, the legal constraints go a long way toward explaining why so many cities failed to resolve the firemen's and policemen's pension problem. In view of the success of recent reform efforts in Washington, Minnesota, and other states, there is reason to believe that the political constraints are less formidable now than a decade ago. Firemen's and policemen's unions are still powerful, but they have been hard pressed to justify their extremely liberal retirement plans in the face of the acute fiscal plight of urban America. At the same time, state pension or retirement commissions have given the reform movement the sustained, sophisticated, and single-minded leadership that is vital for success. There is, however, no reason to believe that the legal constraints are less formidable now than in the past. The contract theory may be preposterous, as Paul Studensky and other members of a New York City committee on pensions put it in 1943.[49] But the judges have given no sign that they are inclined to abandon it; nor have the legislators. And in New York and several other states it would take a constitutional amendment, the likelihood of which is very slim, to nullify it. Thus, even if the political constraints on pension reform are overcome, it will be a long while before the cities are free of the heavy burden of their firemen's and policemen's pension systems.

CHAPTER SIX

Proposals for Change

Despite formidable political and legal constraints, reformers have managed to change big-city firemen's and policemen's pension plans in a number of salutary ways. They prevailed on the authorities to put the pension plans on a contributory basis. By the 1970s, most firemen and policemen were contributing 5 to 9 percent of salary to their pension funds, which is in line with the contribution rates of most nonuniformed employees. Reformers also persuaded the authorities to put the pension plans on a reserve basis. By the 1970s, all but a handful of big cities were required to appropriate enough money to meet long-term costs and to cap, if not necessarily to amortize, unfunded liabilities. Moreover, at the urging of reformers, who argued that the pension funds were yielding a very low rate of return, the authorities eased traditional restrictions on investing in equity. Although their assets were tied up almost entirely in government securities at the end of World War II, most funds held sizable amounts of common stock by the 1970s.[1] These changes helped to generate the sharp rise in revenues in most big-city firemen's and policemen's pension funds in the past few decades.

In an effort to generate even more revenue in the future, reformers have urged authorities to raise employee contributions, put the remaining pay-as-you-go systems on a reserve basis, and remove restrictions on investing in equity. They have pointed out that uniformed officers do not pay a fair share of their pension bills, let alone the one-half that was recommended by Lewis Meriam and others. With the exception of military personnel, whose retirement plan is wholly noncontributory, no public employees bear a smaller share of their pension costs than big-city firemen and policemen. Reformers have also claimed that the pay-as-you-go approach has long been discredited. A fiscally irresponsible policy, it conceals the long-term costs of public employee pensions, passes them on from one generation of taxpayers to the next, leaves in its wake massive unfunded liabilities, and in time places an intolerable burden on the cities. Reformers have stressed, too, that restrictions on investing in equity are anachronistic. If the restrictions were removed—and if professional advisors were hired and investment decisions were streamlined—the trustees of the pension funds should be able to increase the rate of return without violating their fiduciary responsibilities.[2]

If enacted, these proposals would probably produce a good deal of additional revenue. The big cities have such huge fire and police payrolls that a modest increase in the contribution rate would yield anywhere from several hundred thousand to several million dollars a year. For example, if Los Angeles hiked the contribution rate from 7 to 10 percent, as City Administrative Officer C. Erwin Piper recommended in 1979, employee contributions would jump from $16 million to $23 million a year, or from under 16 to over 22 percent of pension payroll. A shift from a pay-as-you-go to a reserve basis would produce even more revenue. According to actuarial studies—which assumed that the cities would meet long-term costs and amortize unfunded liabilities—such a change would force the authorities to raise their annual contributions from $3 million to $16 million in Pittsburgh and from $6 million to $28 million in Denver. Many big-city firemen's and policemen's pension systems also have such vast assets that a small increase in the rate of return would bring in several million dollars more per year. For instance, if the Detroit fund, with $532 million in assets, raised its rate of return by only 1 percent, its earnings would go up by more than $5 million a year and would climb from 57 to 67 percent of pension payroll.[3]

But the reform proposals would not produce enough additional revenue to solve the pension problem. Most firemen and policemen are already putting 5 to 9 percent of salary into their pension systems. To raise the contribution rate to 10 percent would increase total revenue by very little. To lift it well above 10 percent—which is what would be necessary in most big cities if the uniformed personnel were to bear anything like one-half of pension costs—would probably be regarded as unreasonable. Most firemen's and policemen's pension systems are already operating on a reserve basis, too. If the authorities order the systems to amortize the unfunded liabilities and shorten the amortization period, a good deal more revenue would be required. But in view of the fiscal bind in which most big cities now find themselves, it is hard to see how they could come up with the money. Most firemen's and policemen's pension funds are already investing heavily in equity as well. It might be possible for them to raise the rate of return by putting even more of their assets into the stock market. But it would be risky. The market goes down as well as up. It did well in the 1950s, but it did poorly in the mid-1960s and early and late 1970s.[4] And at some point the desire to increase the yield must give way to the duty to safeguard the assets.

Hence most big-city firemen's and policemen's pension systems will still be in serious trouble even if they raise the contribution rate, operate on a reserve basis, and increase the rate of return. For the pension problem is, at heart, a problem of high costs, not of low revenues. In most systems, revenues have been rising, steadily and sharply, for at least two decades. Take the case of the Los Angeles Fire and Police Pension System. From 1960, when the system went back on a reserve basis, to 1979, the city's contributions climbed from $6 million to $146 million, and the investment earnings soared from $0.5 million to $45 million. With the employee contributions rising from $3 million to $16 million, the system's revenues jumped from $10 million to $207 million. Despite this enormous increase, the system's obligations continued to go up. And its unfunded liabilities soared from $315 million in 1960 to $2.5 billion in 1979—from $40,000 to $210,000 for each uniformed officer and from six to ten times the payroll.[5] As long as Los Angeles and other cities are supporting almost as many retired officers and survivors as active officers, and supporting most of them fairly well, the pension problem cannot be solved by raising revenues. It can only be solved by cutting costs.

The reformers have long been after the authorities to cut costs by reducing (or, at any rate, holding down) the pension rolls and the retirement allowances. But until recently the results of their efforts were, for the most part, disappointing. While reformers were successful in Detroit in the late 1960s, they were unsuccessful in Los Angeles in the mid-1960s. And they fared no better in Minnesota, where firemen and policemen blocked the state retirement commission's proposal to tighten eligibility requirements and lower retirement benefits in the late 1960s. In a few places, the authorities even gave in to pressure from the uniformed forces to liberalize the pension systems. Congress eliminated the age requirement for District of Columbia firemen and policemen in 1970, allowing them to retire at any age after twenty years of service. Four years later, San Francisco voters adopted a cost-of-living escalator that pegged the retired officers' allowances to fire and police salaries.[6]

By the mid-1970s, however, the situation began to change. Troubled by the mounting deficits of the Law Enforcement Officers and Fire Fighters Retirement System (LEOFF)—the statewide firemen's and policemen's pension system that was set up in 1969—the Washington legislature passed a law in 1977 that raised the normal retirement age from fifty to fifty-eight, reduced the death and disability benefits, and placed a 3 percent ceiling on the cost-of-living escalator. Two years later, Congress revised the District of Columbia firemen's and policemen's pension system, hiking the service requirement from twenty to twenty-five years, adding an age requirement of fifty, lowering the retirement allowances, and tightening the requirements for disability pensions. Late in the decade, the Colorado legislature closed the Denver firemen's and policemen's pension systems to new appointees, who were required to join a statewide retirement plan that imposed stiffer eligibility requirements, gave smaller retirement allowances, and provided less generous postretirement increments. Perhaps the most remarkable turnabout took place in Los Angeles, where, after more than half a century of attempts at pension reform, the voters approved a charter amendment in 1980 that imposed an age requirement of fifty, reduced retirement allowances for officers' widows, and put a 3 percent ceiling on the cost-of-living escalator.[7]

The prospects for pension reform are brighter now than ever. Most firemen's and policemen's pension systems are in precarious condition; and in the wake of the fiscal crises in New York and Cleveland, the warnings that a "Day of Reckoning" is at hand cannot be ignored. Hamstrung by California's Proposition 13 and other tax-limitation and tax-reduction measures, the cities will find it difficult to bail the pension systems out. They can expect little help from the states and even less from the federal government. While the firemen's and policemen's unions will continue to oppose efforts to tighten eligibility requirements and lower retirement benefits, they are less powerful now than a decade ago. As long as many cities are having trouble paying their bills, the unions find it hard to justify what are by far the most liberal pension plans around. This is, therefore, a good time to analyze some of the principal proposals that are currently being put forth to cut firemen's and policemen's pension costs, to consider whether these proposals will reduce (or, at any rate, hold down) the pension rolls and the retirement allowances, and, if it turns out that they will, to figure out what it will cost the uniformed officers and the citizens who rely on them.

One of the reformers' principal proposals is to tighten the eligibility requirements for service pensions—above all to impose an age requirement of at least fifty. This proposal has long been high on the reform agenda. The New York City Commission on Pensions called for an age requirement of fifty-eight in 1918. The Los Angeles Municipal League suggested an age requirement of fifty in the mid-1920s; and several years later the Los Angeles Realty Board recommended a requirement of sixty. The Detroit Pension Commission proposed an age requirement of fifty-five in the mid-1930s, and the New Orleans Bureau of Governmental Research recommended an age requirement of sixty in 1944.[8]

Despite many setbacks, reformers pressed ahead after World War II. Starting in the late 1960s, their campaign met with more success. At the behest of a special committee appointed by New Orleans Mayor Victor H. Schiro, the Louisiana legislature imposed an age requirement of fifty in 1967. A year later, the Cavanagh administration prevailed on the voters to approve a requirement of fifty-five. Congress, which had abolished a longstanding requirement of fifty in the District of Columbia in 1970, reimposed it nine years later. And Los Angeles gave its approval to an age requirement of fifty in 1980.[9] By then, most big cities had an age requirement, and many others were under pressure to fall in line.

Reformers have stressed that an age requirement would prevent firemen and policemen from retiring at what the New Orleans Bureau of Governmental Research once referred to as "absurdly early ages," thereby slowing down the growth of the pension rolls and helping the uniformed services retain veteran officers in "the prime of life." On the surface, this argument is quite plausible. As things now stand, most firemen and policemen enter the service in their early or middle twenties, work for about twenty-five years, and, if they opt for a service pension, retire at about fifty. Assuming normal life expectancy, they spend about as much time on the pension rolls as on the job (and even more if their survivors are taken into account). But if an age requirement of fifty-five were imposed, the argument goes, most uniformed officers would spend more than three years on the job for every two years on the rolls. If the requirement were set at sixty, they would spend more than two years on the force for every year on retirement, which was more or less the practice in the early twentieth century. Even an age requirement of fifty would make a significant difference in such cities as Los Angeles, where more than one out of every three policemen and nearly one out of every six firemen who retired on a service pension in the 1970s was in his forties when he left the service.[10]

But an age requirement is open to question on the grounds that it would undermine the effectiveness of the uniformed services. According to this view, policing and fire fighting are, as a high-ranking federal official put it, "a young man's work." The jobs are so hazardous and strenuous that after fifteen or twenty years most officers are, to quote the *Firemen's Journal*, "pretty well battered up and worn out." The American people are aware of this; as Edward J. Kiernan, president of the International Conference of Police Associations, has pointed out, "They do not want a 60-year-old cop chasing a bank robber down the street." They do not expect a man of fifty-five to fight fires any more than they expect him to play professional football, observed David A. Ryan, president of Local 36, International Association of Fire Fighters. An age requirement would also fill the departments with many overaged and worn-out officers who can no longer do their jobs, spokesmen for the firemen and policemen have insisted. And as Joseph S. Goldring, president of the Police Association of the District of Columbia, testified, the requirement would discourage qualified young men from joining the service and thereby weaken efforts to upgrade the uniformed forces.[11]

An age requirement is also open to question on the grounds that it would not necessarily slow down the growth of the pension rolls. As Donald Randall, president of a Washington, D.C., police union, pointed out, many officers will circumvent the requirement by going out on a disability pension. Although the relationship between service pensions and disability pensions is complicated, there is some evidence to suggest that Randall may be right. Between 1908 and 1914, a time when New York City's police pension fund had an age requirement of fifty-five, seven out of every ten policemen went out on disability—which was twice as high as the rate for firemen, who were eligible for a service pension after twenty years regardless of age. Between 1925 and 1932, by which time the age requirement was eliminated, only one out of every ten policemen retired for disability. In the late 1960s, when the Washington, D.C., firemen's and policemen's pension plan had an age requirement of fifty, about 95 percent of the officers went out on disability. But after 1970, when Congress abolished the age requirement (and tightened the processing of applications), the disability rate dropped sharply; and by 1977, it was down to 55 percent.[12]

While it is not clear that an age requirement would undermine the effectiveness of the uniformed services, there is a good chance that it would raise the fire and police disability rates. This is a grave matter. The rates are already very high—so high, in the words of Dan M. McGill, chairman of the Wharton School's Pension Research Council, that "it's almost a national scandal." The rates are at their highest in Washington, D.C., where in 1977 close to five out of every six retired officers were on the disability rolls, a ratio that led Senator Thomas F. Eagleton to brand the District's pension system as "far and away the premier ripoff system" in the nation. But they are also very high elsewhere. In New York City, where Philip R. Michael, chairman of the Mayor's Committee on Pensions, remarked that things have "gotten out of hand," more than one out of every two retired firemen and policemen was receiving a disability pension in 1978. In San Francisco, where the *Examiner* quipped that firemen's and policemen's disability pensions were "the biggest giveaway since supermarket bingo," roughly one out of every two retired officers was on the disability rolls in 1979. In the state of Washington, more than six out of every ten retired officers went out on disability in the first six years of the LEOFF system's operations, a rate that "staggers the imagination," observed Thomas P. Bleakney, a West Coast actuary.[13]

Some cities have fairly low fire and police disability rates. Of every one hundred officers on the pension rolls in 1978, eight went out on disability in Chicago, twelve in Detroit, and seventeen in Pittsburgh. But these cities were the exception. One out of every five retired officers was on the disability rolls in Los Angeles and one out of every three in St. Louis. Roughly six out of every ten retired officers went out on disability in Baltimore, and nearly two out of every three in Denver. Moreover, the disability rates were much higher for firemen and policemen than for nonuniformed employees in all but a handful of cities. The rates were twice as high in Los Angeles, three times as high in Baltimore, five times as high in Minneapolis, and even higher in St. Louis, New York City, and Denver.[14]

Not only are the fire and police disability rates very high, but in most cities they have risen sharply in the past two decades. In Portland, the disability rate went up from 13 percent in 1957 to 37 percent in 1972, before dropping to 31 percent in 1978. In St. Louis, so many firemen went out on disability in the mid- and late 1970s that the rate rose from 22 percent in 1957 to 33 percent twenty-one years later. The disability rate went up much less in Minneapolis, Chicago, Denver, and Atlanta. And in Detroit the disability rate declined slightly. But elsewhere, the fire and police disability rates went up much more. Between 1957 and 1978, they climbed from 1 to 17 percent in Pittsburgh, from 14 to 20 percent in Los Angeles, and from 25 to 33 percent in Oakland. In New Orleans, so many firemen went out on disability that the number of retired officers on the disability rolls soared from less than one out of every forty in 1957 to more than one out of every four twenty years later.[15]

Lately, fire and police disability rates have fallen in a few cities and gone up very little in others. Between 1972 and 1978, for example, they dropped from 24 to 15 percent in Indianapolis and went up just three points in Denver. And in Washington, D.C., the number of officers retiring on disability fell from over nine out of ten in the late 1960s to under six out of ten in the mid-1970s.[16]

But in most cities, "the trend is up," according to M. Lewis Thompson, former head of the Los Angeles Fire and Police Pension System. In Baltimore, roughly ten times as many firemen and policemen went out on disability in the mid-1970s as the actuaries had predicted—an "extraordinary" increase, said Janet L. Hoffman, the City Council's fiscal advisor. Chicago's disability rate—among the lowest in the nation—climbed from less than 3 percent in 1972 to more than 8 percent in 1978; and Pittsburgh's jumped from less than 3 percent to more than 17 percent. At the same time, the rates went up roughly 50 percent in Oakland and St. Louis.[17]

The fire and police disability rates are so high partly because many officers have applied for disability pensions and partly because most of the applications—close to 90 percent in San Francisco and even more in Washington, D.C.—have been approved by the authorities. The flood of applications is due to more than just the hazards of policing and fire fighting, which spokesmen for the uniformed forces claim were worse than ever in the 1960s and 1970s. By far the most important reason is that disability pensions are generally larger than service pensions. For example, if a St. Louis fireman retires after twenty-five years of service, he receives half of the average salary in his last three years; if he goes out on a service-connected disability, he gets three-quarters of his final salary. If a San Francisco policeman takes a service pension, his widow is entitled to half of his allowance; if he is given a disability pension, she is entitled to three-quarters. Moreover, if an officer's disability is service-connected, and the great majority of them are, his allowance is exempt from federal income tax. Thus he earns almost as much on the pension rolls as he did on the force—and even more if he finds another job. The disability benefits are "a real pot of gold," remarked Foster B. Roser, personnel director of Philadelphia, where six out of every seven policemen who retired in 1971 went out on a service-connected disability.[18]

Many officers who have applied for a disability pension have probably been influenced by the liberal definition of disability that has gained widespread acceptance in recent years, a definition that helps to account for the rising disability rates among federal civil servants, nonuniformed municipal employees, and Social Security recipients. But some officers have applied for a disability pension for reasons that are unique to policing and fire fighting. Many firemen's and policemen's pension plans are not vested; their members are not entitled to a pension (or any part thereof) until they meet all the eligibility requirements. For a fireman or policeman who is fed up after five or ten years, a disability pension is the only way to leave the service without forfeiting his benefits. Other officers ask for a disability pension in order to avoid facing charges of corruption or brutality. They are often supported by their superiors, who want to spare the department unnecessary embarrassment. Still other officers file for a disability pension under pressure from headquarters—a point made by Baltimore officials who accused Commissioner Donald D. Pomerleau of using the pension system as a "dumping ground" for officers who did not fit into his plans for upgrading the police department.[19]

If a good many applicants are not disabled, much less disabled in the line of duty, why do the authorities approve the great majority of applications for a disability pension? The answer lies partly in the composition of the retirement boards, which study the officers' claims, review the medical reports, and then make the decisions. Firemen and policemen are well represented on most of these boards; in 40 percent of the pension systems they hold over half the seats. The arrangement in Seattle—where nearly four out of every ten policemen and more than six out of every seven firemen went out on disability under the LEOFF plan in the 1970s—is typical. Firemen elect two officers to a five-man board, and policemen choose three members of a seven-man board. Sympathetic to their fellow officers and accountable to them come election day, the employee representatives usually support applications for a disability pension. "[I lean] over backwards a little," admitted Leon Bruschera, secretary of the San Francisco firemen's union, who in one six-month span voted in favor of eighty-eight out of eighty-nine such applications from fire fighters.[20]

Even if the firemen and policemen were not well represented on the retirement boards, most applications for disability pensions would probably be approved anyway. To begin with, many applicants claim to have back and neck injuries, which are very hard to diagnose. Even harder to diagnose are the claims of mental stress, which have gone up sharply in recent years. In the absence of clear-cut medical evidence to the contrary, many boards are reluctant to turn down these applications. What is more, many applicants have heart or lung ailments; unless they can prove otherwise, many boards are required to treat these disorders as service-connected.

According to the *New York Times*, there was a "virtual epidemic" of service-connected disabilities among the city's uniformed officers after passage of the state's heart and lung law in 1970. Another reason some retirement boards find it extremely hard to turn down applications for disability pensions is that most fire and police chiefs have taken the position that the departments have no light duty, that they have no place for an officer who is less than 100 percent fit. Still another reason is that many courts have ruled that, in view of their humanitarian purpose, the retirement laws should be "construed liberally," giving the applicants the benefit of the doubt in the event of conflicting or ambiguous medical reports.[21]

Reformers have put forward a host of proposals to reduce fire and police disability rates. Some are designed to make it less attractive for the officers to apply for a disability pension. One such proposal is to bring disability benefits down to the same level as service benefits. Closely related is a proposal to base the retirement allowance on the degree of disability, a practice that is followed in the administration of Workmen's Compensation. Other proposals are intended to make it less difficult for the boards to turn down an application. One such proposal is to assign cases to independent hearing officers, who would presumably be less susceptible to political pressures than retirement board members. Even more significant is a proposal to eliminate the heart and lung presumptions and compel officers to prove their injuries were incurred in the line of duty. Still other proposals are designed to get officers who are no longer disabled off the pension rolls. To this end, reformers have urged retirement boards to require retired officers to undergo periodic checkups and to impose strict limits on outside earnings. They have also pressed the fire and police departments to provide light duty for officers who are not 100 percent fit but not totally disabled either.[22]

Some of these proposals have been put into practice. As part of an overhaul of the statewide LEOFF system, Washington recently brought disability allowances down to the same level as service allowances. Instead of an automatic 50 to 60 percent of final salary, disabled officers now get 2 percent of final average salary (defined as highest salary in five consecutive years) for each year of service. As part of a long and fairly successful effort to hold down disability rates, Chicago required that retired officers be examined by a doctor at least once a year to see if they are well enough to return to active duty. Extensive steps have been taken in Washington, D.C., where Congress recently abolished the aggravation clause, tightened the limits on outside earnings, and mandated annual, rather than biennial, medical checkups. In the case of new appointees, it went even further; instead of an automatic two-thirds of final salary, disabled officers now get an allowance based on degree of disability, with a minimum of 40 percent of final salary. Along the same lines, the Washington, D.C., police department set up a Casualty Investigation Unit—a thirteen-man undercover squad, equipped with vans and cameras—that travels around the country looking into alleged abuses of the District's disability program.[23]

Even assuming that the reform proposals were widely adopted, which is by no means a sure thing, it is hard to say whether they would work out as expected. If they did, the imposition of an age requirement for service pensions would no doubt improve the financial condition of the big-city firemen's and policemen's pension systems. But if they did not, the requirement would probably drive up disability rates, which would have serious fiscal repercussions. One is that the average retirement allowance would go up. Most officers receive a higher allowance if they go out on disability (or, at any rate, on service-connected disability) than if they retire on a service pension. In New York City, for example, the firemen and policemen who were disabled in the line of duty received on average more than $14,000 in 1979, roughly half again as much as the officers who went out on a service pension. Another repercussion is that employee contributions would go down. In most cities, disability pensioners usually spend fewer years on the force than service pensioners and therefore put less into the pension funds. In Los Angeles, for example, the firemen and policemen who were on the disability rolls in 1979 had worked only sixteen years before retiring—about ten years less than officers who left with a service pension.[24]

By far the most serious repercussions of a rise in the disability rates is that the pension rolls would grow at an even faster clip. For disabled officers retire earlier than service pensioners and, as their life expectancy is much the same, spend longer on the rolls. The disparity is striking in Los Angeles. Between 1970 and 1979, the firemen and policemen who went out on disability were on average forty-three years old, or ten years younger than the officers who retired on a service pension. Assuming normal life expectancy, the disabled officers will remain on the pension rolls for thirty years, or half again as long as the service pensioners. The disparity was less striking in Washington, D.C., and New York City. But it was even more striking in Oakland, where firemen and policemen on the disability rolls in 1975 retired on average at forty-six—eleven years earlier than officers on the service rolls.[25]

Another of the reformers' principal proposals is to integrate the firemen's and policemen's pension plans with the Social Security system. This proposal is closely linked to the ongoing campaign for universal Social Security coverage, a campaign that has focused largely on public employees, who were excluded from the system in 1935 for legal and political reasons. At the prodding of the Advisory Council on Social Security, which first advocated mandatory coverage for public employees in the late 1930s, Congress has held hearings on the subject on several occasions. The campaign has gained the support of Robert Tilove, a prominent pension consultant; Robert J. Myers, former chief actuary of the Social Security Administration; and John Bragg, chairman of the public pension task force of the National Conference of State Legislatures. The President's Commission on Pension Policy has called for mandatory coverage for new employees, and New York State's Permanent Commission on Public Employee Pension and Retirement Systems has made integration with Social Security the cornerstone of its reform efforts.[26]

In defense of universal coverage, reformers have stressed that some employees do not have adequate protection—either because they quit their jobs before their retirement benefits are fully vested or because these benefits are inadequate in one way or another. Other employees, who are covered by liberal pension plans, obtain Social Security credits by moonlighting, a practice that provides them with a windfall on retirement. In the absence of mandatory coverage, yet other employees, some of whom earn a high salary, do not bear a fair share of the burden of financing the retirement benefits of less well-off Americans.[27]

Reformers have also pointed out that the adoption of universal coverage, especially the inclusion of the more than two million federal civil servants who do not participate in Social Security, would provide the system a sizable amount of extra money and help to stave off what many regard as an impending fiscal crisis. Reformers have noted as well that if Social Security benefits were used to offset retirement allowances—that is, if the retirement allowances were reduced by all or part of the Social Security benefits—the pension systems would pay out much less in benefits. Even taking into account the employers' share of Social Security contributions, the cities would save quite a bit of money.

Opposition to mandatory coverage has come from several sources. These include public employee unions, which have argued that if their members were covered by Social Security, state and local governments might abolish their pension plans; the National Conference of State Legislatures and the National Governors' Association, which have been reluctant to relinquish additional authority to the federal government; and the Advisory Commission on Intergovernmental Relations (ACIR). Mandatory coverage, the ACIR has insisted, would raise public sector labor costs. Not only would state and local governments have to make sizable contributions to Social Security, but in many places they would not be able to reduce retirement benefits for current employees. If designed to supplement, rather than to offset, pension benefits, integration with Social Security would boost retirement incomes to excessive levels, particularly in the case of firemen and policemen. If mandatory coverage were extended only to new employees, many governments would have to set up separate plans for them, a move that would raise administrative costs and create morale problems. Social Security may well need relief, the ACIR has acknowledged, but it "should not come at the expense of state and local governments and their employees."[28]

Although Congress has not yet adopted mandatory coverage for state and local employees, it has long abandoned the policy of excluding them. The first break took place in 1950, when Congress opened the system to public employees who were not covered by state or local pension plans. Four years later, it allowed covered employees to join as a group, but only if a majority of the group voted in favor of doing so. Shortly after, Congress eased the requirements again, giving current employees in certain states the option of signing up as individuals. So many public employees took advantage of these arrangements that by the mid-1970s 72 percent of the full-time state and local work force were covered by Social Security. Apart from federal civil servants, who have steadfastly opposed mandatory coverage and are now the largest block of uncovered employees, firemen and policemen are the most conspicuous holdouts. At their request, they were excluded from the 1954 legislation. And although Congress later opened Social Security to some firemen and policemen, most have chosen not to join. Thus by the mid-1970s, only about one-third of the nation's uniformed officers were covered by Social Security.[29] Outside of New York and Buffalo, where the officers get Social Security on top of their retirement allowances, very few of the big-city firemen and policemen are covered by Social Security.

Firemen and policemen have strongly opposed efforts to force them into the Social Security system on several grounds. Integration would lead to the abolition of their pension plans, their spokesmen told the Senate Finance Committee in 1954; the public would not support two separate retirement systems. While Social Security is appropriate for the industrial work force, for which it was originally designed, it is not suitable for the uniformed services. Repeating the arguments advanced in the late nineteenth century, the spokesmen pointed out that policing and fire fighting are a young man's work. Under Social Security, however, a member cannot ordinarily collect benefits until he reaches sixty-five (or, at a minimum, sixty-two). Thus, if uniformed officers were forced into Social Security, they would have to stay on the job long after they could do the work—a practice that would turn police stations into "rest homes," a Detroit union leader remarked in 1954.[30] By making it harder to get rid of superannuated officers, integration would lower the caliber of the uniformed services, reduce their efficiency, and undermine their capacity to cope with crime and conflagration.

Firemen and policemen have strongly opposed Social Security coverage for reasons other than concern for public safety. One, which their leaders have made explicit, is that most officers do not believe they can afford to belong to both the Social Security system and a local pension plan. If forced into Social Security, most officers would have to shell out more than 12 percent of salary for retirement. Many would have to put in more than 15 percent. With another 25 to 35 percent withheld for federal, state, and local taxes, many officers could not make ends meet, Edward J. Kiernan, president of the International Conference of Police Associations, pointed out in 1976. Another reason, which the leadership seldom talks about, is that most officers pick up enough credits (through military service, moonlighting, and/or postretirement employment) to qualify for Social Security anyway. Moreover, an officer who retired in 1960 and then worked for twelve years in covered employment would get fully 70 percent of the Social Security benefits he would have gotten if he had been covered while on the force. And he would have put in less than half as much money.[31] Most officers see no reason to pay for benefits that they can get without paying for them, especially if these benefits are to be offset against their retirement allowances.

Although the opposition to Social Security is more than a little self-serving, it has raised at least one objection that cannot be lightly dismissed. Most firemen's and policemen's pension plans have long allowed officers to retire in their late forties or early fifties. A corollary of the assumption that the uniformed services are quasi-military occupations that require young men in prime condition, this policy still commands much support today. But under ordinary circumstances, the Social Security system does not pay retirement benefits until a member reaches sixty-five (or, at a reduced level, sixty-two). And in an attempt to deal with the system's fiscal problems, the Reagan administration has proposed raising the age requirement to sixty-eight. To integrate the plans is by no means an easy task. As Ivan Eldher, secretary-treasurer of the Denver Policemen's Protective Association, told the Senate Finance Committee in 1954, the average officer joins the service at twenty-seven and is eligible to retire at fifty-two. If the plan is integrated with Social Security, Eldher asked, "What will he do during the time he has to wait for social security in order to get full coverage at 65[?]" He cannot afford to retire on partial benefits; and if he remains on the job until sixty-five, which is highly likely, he cannot do the work expected of him.[32]

Recent events in New York State reveal how hard it is to integrate the firemen's and policemen's pension plans with the Social Security system. At the urging of the Permanent Commission on Public Employee Pension and Retirement Systems, which launched a campaign for pension reform in the early 1970s, the state legislature set up a new retirement plan for public employees in 1976. Known as CO-ESC, for Coordinated Escalator, the plan reduced the employees' retirement allowances by half their Social Security benefits and, as a concession to the employee groups, added a 3 percent cost-of-living escalator. Covered by the plan were New York City's firemen and policemen, who had opted to participate in Social Security in the mid-1950s because of its survivors' benefits and who thereafter received Social Security benefits on top of their retirement allowances. Under CO-ESC an officer was allowed to retire after twenty-five years of service at any age; the escalator clause would start when he retired, while the offset provision would begin when he reached sixty-two. The legislature, however, had stipulated that the new plan would not apply to uniformed officers if it cost more than 1 percent of payroll more than the old plan. The actuaries soon reported that it did.[33] With most officers retiring at about fifty, it turned out that the escalator clause would raise expenses more than the offset provision would lower them. Hence the city's firemen's and policemen's pension plans are not yet integrated with Social Security, and many retired officers still receive Social Security benefits on top of their retirement allowances.

Another of the reformers' principal proposals is to abolish the full cost-of-living escalators (or, at the very least, to put a cap on them). Reformers have opposed escalator clauses since the late 1920s, when the Chicago Civic Federation strongly objected to a proposed revision of the firemen's pension system that would have provided postretirement increases pegged to current salaries. Despite their opposition, many cities established cost-of-living escalators after World War II; and few got rid of them. Starting in the late 1960s, however, the situation began to change. At the urging of the Cavanagh administration, Detroit imposed a 2 percent cap on the cost-of-living escalator in 1968; San Francisco did the same about eight years later. Not long after, Washington limited the LEOFF system's escalator to 3 percent; Los Angeles followed suit in 1980. And at about the same time, Minnesota passed a law that abolished the automatic cost-of-living escalator in Minneapolis and St. Paul.[34]

There is much to be said in favor of abolishing or capping full cost-of-living escalators. To begin with, they are expensive. Take the case of a public employee who retires at sixty-five on a fully indexed pension of $10,000 a year and, assuming normal life expectancy, lives another thirteen years. If the consumer price index (CPI) goes up 3 percent a year, his cost-of-living increments will come to $26,000, or 20 percent of his total unadjusted benefits. At 5 percent a year, they will add up to $47,000, an additional 36 percent; and at 10 percent a year, they will reach $115,000, an extra 89 percent. At anything like the rate of inflation in recent years, there is no way that the cities can afford to provide their employees complete protection against inflation. Moreover, full cost-of-living escalators are inequitable. According to the Pension Task

Force of the House Committee on Education and Labor, 61 percent of state and local employees (and 63 percent of firemen and policemen) do not have automatic cost-of-living escalators of any kind. Only 4 percent of nonuniformed employees (and 6 percent of firemen and policemen) have a full escalator.[35] Even if the retired employees whose allowances are pegged to wage hikes are added in, no more than 5 percent of the state and local work force (and 17 percent of firemen and policemen) are protected in full against inflation.

But there is also much to be said against abolishing or capping full cost-of-living escalators. A fixed pension depreciates at a rapid rate even in a period of moderate inflation. If costs go up 5 percent a year, an employee's retirement allowance loses 22 percent of its purchasing power in five years, 39 percent in ten, and 52 percent in fifteen. In a period of high inflation, the loss is far worse. If costs rise 10 percent a year, the real value of the allowance falls 38 percent in five years, 61 percent in ten, and 76 percent in fifteen. At double-digit inflation, or even at a rate approaching it, a retired employee on a fixed pension will soon be unable to make ends meet and in time will end up destitute. Moreover, several major pension plans do provide a full cost-of-living escalator. The federal civil service retirement system pegs benefits to the consumer price index (CPI), as does the government's retirement plan for military officers. And though few private pension plans include a full cost-of-living escalator, the Social Security system ties allowances to the CPI.[36] If public pension plans are supposed to provide income security for life, what is unfair is not that some public employees are fully protected against inflation but that most of them are not.

It is hard enough to provide nonuniformed employees a hedge against inflation at a price that taxpayers can afford. In the case of firemen and policemen, it is even harder. In the first place, most uniformed officers retire in their late forties or early fifties, about fifteen years earlier than most nonuniformed employees. Assuming normal life expectancy, they stay on the pension rolls for an average of nearly twenty-five years, which is close to twice as long as nonuniformed employees. Hence their retirement allowances depreciate sharply. If costs go up 3 percent a year, they lose 52 percent of their purchasing power in twenty-five years; at 5 percent a year, the real value falls 70 percent; at 10 percent a year, it drops 91 percent.[37] Take the case of a Chicago policeman who joined the department in the early 1950s at twenty-five, put in twenty-five years, went out at fifty on a service pension of $10,000, and lives another twenty-five years. If costs go up 3 percent a year, the $10,000 allowance will be worth only $4,800 in his last year; at 5 percent, it will be worth $3,000; at 10 percent a year, it will be worth $900.

In the second place, few firemen and policemen are covered by Social Security, which sets them apart from most public employees. While this arrangement reflects the strong preference of the uniformed forces, it poses a serious problem for state and local authorities. In the case of most nonuniformed public employees, the authorities can assume that Social Security benefits will go up with the CPI, provided that the index rises at least 3 percent a year.[38] They can assume that a good share of the employees' retirement income is fully protected against inflation. In the case of most uniformed officers, however, the authorities cannot make this assumption. While they are aware that many firemen and policemen may qualify for Social Security through military service, moonlighting, or postretirement employment, they cannot be sure how many (or, for that matter, which ones). They have to assume that inflation threatens the entire retirement income of many firemen and policemen, and that, in the absence of a full cost-of-living escalator, some officers and their widows will become impoverished.

But to provide firemen and policemen protection against inflation at a price that the cities can afford is easier said than done. As expensive as a full cost-of-living escalator is for nonuniformed employees, it is even more expensive for uniformed officers. Take the case of a policeman who retires on a fully indexed pension of $10,000 and lives another twenty-five years. If the CPI goes up 3 percent a year, his cost-of-living increments will come to $115,000, or 46 percent of his total unadjusted benefits; at 5 percent a year, they will add up to $227,000, an additional 92 percent; and at 10 percent a year, they will reach $733,000, an extra 293 percent.[39]

A full escalator is so expensive because uniformed personnel retire so early. If the CPI climbs 3 percent a year, a full escalator will cost three times as much for a fireman who retires at fifty as for a bookkeeper who retires on the same pension at sixty-five; at 5 percent a year, it will cost four times as much, and at 10 percent a year, five times as much. A full escalator is also very expensive because most uniformed personnel are not covered by Social Security, the benefits of which are fully indexed. In the case of most firemen and policemen, an escalator has to apply to the entire retirement allowance—and not, as in the case of most nonuniformed employees, only to that part of it that is not offset by Social Security benefits.

If the rate of inflation were low enough, a partial cost-of-living escalator might provide firemen and policemen a hedge against inflation at a price that cities could afford. Given a 4 percent a year increase in the CPI and a 2 percent a year escalator, retirement allowances would lose 37 percent of their value in twenty-five years, and pension costs would rise 28 percent. At a rate of inflation much above 4 percent, however, a partial escalator would give firemen and policemen too little and cost cities too much. Given a 6 percent-a-year hike in the CPI, a 2 percent escalator would let allowances depreciate 60 percent, and a 3 percent escalator would let allowances depreciate 46 percent. At 8 or 10 percent-a-year inflation, a 2 or 3 percent escalator would leave retired officers destitute. Even if the escalator were pegged to half the increase in the CPI, allowances would fall 60 percent if the index rose 8 percent a year and 70 percent if it went up 10 percent a year. And the price would be staggering. At 8 percent, the escalator would raise costs 67 percent; at 10 percent, it would drive them up 91 percent. There simply is no way that cities can provide retired officers lifetime security at reasonable cost at the rate of inflation in the late 1970s.

It can be argued that this is the officers' problem, not the cities'—that if the cities cannot afford more than a 2 or 3 percent escalator, retirees will have to get by on it. In light of the fiscal plight of most firemen's and policemen's pension systems, the argument makes some sense. But it should be kept in mind that, if cities abolish or cap the full cost-of-living escalators, they cannot claim to provide officers lifetime security. Moreover, in the absence of a full escalator, firemen and policemen may choose to retire as early as possible—a strategy that enhances their prospects of obtaining another job.[40] If many do so, it will foster the growth of the pension rolls and deprive the uniformed forces of many veteran officers. Other firemen and policemen may choose to retire as late as possible—a strategy that would increase retirement allowances and reduce the number of years over which they depreciate. If enough do so, it will delay the retirement of superannuated officers, block the appointment of qualified recruits, and perhaps undermine the efficiency of the uniformed services.

In an attempt to solve the firemen's and policemen's pension problem, some reformers have proposed a more drastic measure than either tightening the eligibility requirements, integrating the plans with Social Security, or putting a cap on the cost-of-living escalators. Convinced that the problem cannot be solved as long as the cities retain control, they have suggested that the states replace local firemen's and policemen's pension systems with a consolidated statewide retirement plan for uniformed personnel. If consolidation is not feasible, they have recommended that the authorities close existing systems to new appointees and require them to join a statewide plan. The New Jersey Advisory Commission on Local Policemen and Firemen's Pension Funds came out in favor of statewide consolidation, as did the Ohio Study Committee on Police and Firemen's Relief and Pension Funds and the Minnesota Legislative Commission on Pensions and Retirement. Statewide consolidation was also endorsed by the Association of Washington Cities, the New Orleans Bureau of Governmental Research, and the Colorado Municipal League.[41]

The movement for statewide consolidation goes back at least to the 1930s, when Rochester, Syracuse, and Buffalo closed their firemen's and policemen's pension plans to new appointees and compelled them to join the New York State Retirement System.[42] But it made little progress before World War II. To begin with, statewide consolidation was not high on the reformers' agenda. The reformers saw no reason that big-city firemen and policemen should have separate pension plans, much less separate plans with such handsome benefits. But if they favored consolidation as an alternative, it was citywide, not statewide, consolidation. They wanted to put uniformed officers into citywide pension systems for non-uniformed employees, not into statewide retirement plans for public safety personnel. On the advice of George Buck, Boston, Cincinnati, and San Francisco all opted for citywide consolidation in the 1920s and 1930s.

What is more, state takeover was not appealing to the big cities. As far as municipal officials were concerned, the firemen's and policemen's pension plans were in trouble, but not in serious trouble. Most officials were confident that they could deal with the problem and were disinclined to call on the state to bail them out. They were reluctant to relinquish authority to the state legislators and to give up power over the processing of disability applications, the investing of pension fund assets, and other matters that could be exploited to enhance their political influence.

To make matters worse, big-city firemen and policemen were sharply opposed to statewide consolidation. They feared that it would not only deprive them of a say in the management of their pension plans but also lead to tighter eligibility requirements, lower retirement allowances, and higher contribution rates. Well organized and well connected, the uniformed officers lobbied hard to head off consolidation. To give one example, the New York State Pension Commission sponsored a bill in 1931 that would have replaced local pension plans with a statewide retirement system. New York City's firemen and policemen strongly objected; and with what one of their spokesmen called "the co-operation of our upstate brothers," they defeated the bill.[43]

If all this were not enough, the states were anything but enthusiastic about taking responsibility for local firemen's and policemen's pension systems. Although many states set up retirement plans for their own employees in the 1920s and 1930s, most legislators felt that the states had no obligation to provide pensions for municipal employees. That was the cities' responsibility. In some states, the legislatures were willing to establish statewide retirement plans for municipal employees who were not covered by local pension systems. But nowhere were they eager to take over big-city firemen's and policemen's pension systems—especially if it meant assuming all or part of the enormous deficits that had built up in the first third of the century.

After World War II, however, the movement for statewide consolidation picked up momentum. One reason is that reformers lost faith that citywide consolidation could solve the pension problem and gradually came to the conclusion that statewide consolidation was the only hope. As John Mandeville, executive secretary of the Minnesota Legislative Commission on Pensions and Retirement, pointed out, consolidation would reduce administrative costs. It would provide uniform benefits, thereby fostering equity among firemen and policemen and discouraging "leapfrogging," the practice by which public employees attempt to outdo one another in wresting concessions from the authorities. It would also allow an officer to move from one fire or police department to another without sacrificing his retirement benefits, an option that is available to very few firemen and policemen.[44] Nor was that all. By making the states pay a large share of the pension bills, consolidation would give them a strong incentive to reduce pension costs. By transferring control from local boards to state agencies—where firemen and policemen, it is commonly assumed, have less clout—it would enhance the prospects of pension reform. Finally, consolidation would shift the pension problem from the cities to the states, which have the financial resources to cope with it.

Another reason the movement gained momentum is that some big cities reversed their position on statewide consolidation. Alarmed by the soaring costs and growing deficits of the firemen's and policemen's pension systems, many municipal officials began to doubt that they could meet their obligations to retired officers without making drastic cuts in vital services. Save for massive infusions of state aid, which were highly unlikely, consolidation seemed to offer the only way out of the morass. Having turned several essential functions over to the states after World War II, the cities had no compunction about asking them to take over the pension systems. Municipal officials justified this request on the grounds that the states were largely to blame for the pension problem, arguing that it was the states, not the cities, that had given the uniformed services their liberal pension plans. Supported by such lobbying groups as the Association of Washington Cities, they also stressed that statewide consolidation would help the small towns. This concern for the well-being of the small towns was often taken at less than face value, especially in Colorado, where the consolidation bill introduced into the state legislature in the late 1970s was referred to as "the Denver bail-out bill."[45]

The big-city firemen and policemen who were the principal opponents of statewide consolidation labored under a number of severe handicaps. Their fraternal and benevolent associations, which had been transformed from pressure groups into de facto labor unions in the late 1950s and early 1960s, lost much of their political clout in the mid- and late 1970s. And many firemen and policemen had doubts about the wisdom of spending very much of their political capital on an all-out battle against consolidation. Despite a strong preference for separate retirement plans, many officers also had deep concerns about the solvency of their pension systems and thought that consolidation might enhance their families' security. Moreover, big-city firemen and policemen could no longer count on small-town officers to help in the struggle against statewide consolidation. During the 1960s and 1970s, many small-town firemen and policemen, most of whom had less liberal pension plans than big-city officers and some of whom had no pension plan at all, began to realize that a statewide retirement system might be in their interest. And in several states, they supported consolidation.[46] The small-town officers' support for consolidation did more than just deprive big-city firemen and policemen of a valuable source of aid. It undercut the argument that consolidation was nothing more than a roundabout way of bailing out the big cities—an argument that had much to be said for it in some states—and made a strong impression on upstate legislators without whose backing consolidation would not have stood much of a chance.

The states also began to look more favorably on taking responsibility for local firemen's and policemen's pension systems. At the behest of hard-strapped cities, some states started to contribute a lot of money to these systems in the 1960s and 1970s. If the state had to pay the bills, many legislators concluded, it might as well run the systems. Other legislators reached the same conclusion for different reasons. Some were swayed by the argument that the states were largely to blame for the pension problem. Others were convinced that it would be easier for the legislature to set up one sound statewide system than to reform a host of defective local plans. Still others were concerned that if the big-city pension systems went broke, the states might have an obligation to make sure that retired officers and their survivors received their benefits. Despite these considerations, the states might still have resisted the pressure to take over local firemen's and policemen's pension systems were it not that they had grown accustomed in the postwar years to supplying a wide range of services that had once been the exclusive responsibility of the cities. If the state's responsibility extended to transportation, education, and welfare, why should it stop short of pensions?

New Jersey was one of the first states to give in to the pressure for consolidation. It set up a statewide retirement plan for uniformed personnel in the mid-1940s and abolished local firemen's and policemen's pension systems several years later. Ohio followed suit in the mid-1960s, Washington in the late 1960s. Louisiana established one statewide plan for municipal police officers in the early 1970s and another for municipal fire fighters later in the decade. Colorado and Minnesota joined the fold in the late 1970s.[47] By the 1980s, statewide retirement plans covered all or some of the uniformed officers in more than a dozen big cities.

Some big cities bucked the trend. New York City runs its own pension plans for uniformed personnel, as do Atlanta, Detroit, and

Los Angeles. Philadelphia operates a retirement system for municipal employees, uniformed and nonuniformed; so do Boston, Milwaukee, and San Francisco. Whether these cities will call on the states to relieve them of the responsibility to provide pensions for firemen and policemen is not clear. Nor is it clear that the legislatures, which have severe financial problems of their own and powerful suburban and rural constituencies to answer to, will oblige the cities. But it is highly likely that henceforth statewide retirement plans will cover a large and growing number of big-city firemen and policemen.

There is much to be said in favor of statewide consolidation, although probably not as much as the reformers claim. Consolidation might well reduce administrative costs, discourage leapfrogging, and enhance intrastate portability. But these changes would not do much to solve the firemen's and policemen's pension problem. Big-city pension systems spend so little on administration that even a sizable cutback would not improve their financial condition very much. For example, if the Los Angeles Fire and Police Pension System reduced administrative costs by one third, it would save less than $500,000 a year, or only 0.5 percent of total expenditures.[48] Also, big-city firemen and policemen play leapfrog not with small-town officers in their own states but with big-city officers in other states. Whether covered by a local or a statewide plan, big-city firemen and policemen will press for parity with whichever big-city officers have the most liberal retirement plans. Moreover, the lack of portability does not contribute much to the pension problem, although it does impose a hardship on some officers. If uniformed officers were free to move from one department to another without forfeiting their retirement benefits, the firemen's and policemen's pension problem would probably be even worse.

Still, statewide consolidation would do a good deal to solve the pension problem. As the reformers claim, it would give the states a strong incentive to reduce pension costs. As matters now stand, the states do not have such an incentive. Although the state legislatures set the eligibility requirements, retirement benefits, and contribution rates for many big-city firemen and policemen, they do not pay the pension bills. In New York, Detroit, Chicago, and a few other cities, they do not even make token contributions. Thus, under pressure from the uniformed forces, most states have been prone to liberalize the pension systems without providing the funds to pay for it, a burden that falls on the cities. Under most statewide retirement systems, however, authority and responsibility are joined. The legislature does not just make policy; it also pays the bills (or at any rate a large share of them). And in most states, the legislators are more concerned about costs when they have to meet them. A good example is Washington, where the legislature established the extremely liberal LEOFF system in 1969. Forced to pour so much money into the system that the state's solvency was in jeopardy, the legislature later tightened the eligibility requirements and took other steps to reduce pension costs.[49]

Consolidation would also enhance the prospects of pension reform, although probably not as much as reformers believe. It would have a strong impact in Los Angeles, where uniformed officers have had a great deal of influence in the City Council. And Oakland's firemen and policemen will have less of a say about pension policy now that all new appointees are required to join a statewide plan for uniformed personnel. But consolidation would not make very much difference in New York City, where firemen and policemen have long had a lot of clout in the state legislature—almost as much as in the City Council. The same is true for the big cities in Washington, where, as Mayor George Johnston of Tacoma said in the late 1970s, the uniformed officers "are the biggest lobbyists in the state."[50] Consolidation will make even less difference in these states if big-city firemen and policemen form effective coalitions with small-town officers.

Consolidation would also shift part of the pension problem from the cities to the states. How large a part depends on the terms of the takeover. In Washington, where the legislature transferred all uniformed personnel into a statewide system, the state not only assumed the unfunded liabilities but also bore the brunt of the future costs. But in Minnesota, the state offered only to help the cities amortize the unfunded liabilities. And in California, the state provided coverage for Oakland's new recruits but accepted no responsibility for the city's obligations to its current and retired officers and their survivors.[51] Despite the recent rash of tax-limitation and tax-reduction measures, the states are, as the reformers claim, in a better fiscal position than the cities to cope with the firemen's and policemen's pension problem. Proposition 13 notwithstanding, California has much greater resources than Oakland. Things may be bad in Ohio, but they are worse in Cleveland. If pension costs force service cutbacks, New Jersey can bear up better than Newark. Much the same can be said of Illinois and other states that have not yet taken over the big-city firemen's and policemen's pension systems.

Consolidation may well help to solve the firemen's and policemen's pension problem, but, as the history of Washington's LEOFF system reveals, it is by no means a panacea. An extremely liberal system that covered all firemen and policemen, LEOFF was set up in 1969. It started out with an unfunded liability of $240 million, or $33,000 for each officer and 3.5 times the payroll. To amortize the liability and to meet long-term costs, the actuaries recommended that the state put 33 percent of payroll into the system in the 1971–72 biennium and 38 percent in the 1974–75 biennium. But the legislature, caught in a fiscal squeeze, appropriated only one-tenth of the recommended contribution. In the meantime, the pension rolls soared, largely because disability retirements went up two to three times as much as anticipated. Wage hikes also went up faster than expected, driving up the retirement allowances. With expenses rising and revenues lagging far behind, the unfunded liability climbed to $612 million by 1975, or $102,000 per officer and 5.4 times payroll, and the state's contribution soared to 53 percent of payroll. Convinced, as Senator Martin J. Durkan put it, that "the state was sold a bill of goods," the legislature completely overhauled LEOFF in 1977, tightening the eligibility requirements and reducing the retirement allowances for new appointees.[52]

The lesson to be drawn from the history of the LEOFF system is that statewide consolidation can make a difference under some conditions but not under others. It did not make much of a difference in Washington largely because the legislature replaced the local plans with a statewide system that imposed only slightly stiffer eligibility requirements, provided virtually the same retirement allowances, and pegged the cost-of-living escalator to the CPI. But consolidation will probably make more of a difference in Minnesota, where the statewide system has much tighter eligibility requirements than the local plans, somewhat lower retirement benefits, and a much less liberal cost-of-living escalator. Whether big-city firemen and policemen are covered by statewide systems or local plans, the authorities have to deal with the same issues. They have to decide whether to continue to treat uniformed officers like military personnel, whether to integrate their plans with Social Security, and whether to give them a hedge against inflation. How these issues are resolved will determine whether statewide consolidation helps to solve the pension problem or just shifts the burden from the cities to the states.

Some municipal officials believe that consolidation offers a way out of the morass even if it just shifts the burden to the states. Although this view is understandable, it is more than a little shortsighted. For better or worse, the well-being of the big cities depends heavily on the well-being of the states. The states not only provide the cities a wide range of vital services, they also give them substantial financial aid, which is more important than ever now that the Reagan administration is cutting back on federal aid to the cities. In the aftermath of Proposition 13, Proposition 2½, and other tax-limitation and tax-reduction measures, however, many states are hard pressed to maintain the current level of aid to the cities, much less to increase it. Thus, if consolidation just shifts the burden from the cities to the states, many legislatures will probably be unable to meet their obligations to firemen, policemen, and their families without making deep cuts in vital services and financial aid. For most big cities, such cuts would be extremely painful.

Even if statewide consolidation helps to solve the pension problem, it has one serious drawback: it reduces the autonomy of the big cities. Consolidation is not an isolated phenomenon. It is part of a longstanding process by which the big cities have relinquished a host of essential functions to state agencies, metropolitan boards, and public authorities. The process went so far in Los Angeles, Mayor Samuel Yorty told a Senate committee in 1966, that he had no authority over housing, health, education, welfare, employment, or transportation.[53] Consolidation would force the cities to turn over still another essential function to the states. Moreover, it could be argued that if the states take over the firemen's and policemen's pension systems, perhaps they should take over the fire and police departments as well. As policing and fire fighting are two principal functions still carried out by the cities, such a takeover would probably destroy much of what little remains of municipal autonomy in America. That many cities are anxious to give up control over firemen's and policemen's pensions is strong evidence of the severity and intractability of the pension problem.

Conclusion

In order to reduce firemen's and policemen's pension costs, several closely related problems must be faced. If most uniformed officers retired in their late fifties and early sixties, it would not be so hard to integrate their pension plans with Social Security. If their plans were integrated with Social Security, their retirement income would not lose so much of its purchasing power and their cost-of-living escalators would not be so expensive. If firemen and policemen received a fixed pension, many of them would not be so antagonistic to Social Security coverage and so enthusiastic about early retirement.

Of these problems, early retirement stands out as far and away the most perplexing. The enduring legacy of the two-tiered pension system that emerged in the late nineteenth and early twentieth centuries, early retirement is the practice that most sets the firemen's and policemen's pension plans apart from the retirement systems for nonuniformed employees. It complicates the prospects of integrating the firemen's and policemen's pension systems with Social Security and of preventing the depreciation of the officers' retirement income. Along with the reduction in entry age, imposition of the merit system, and increase in life expectancy, early retirement also stimulates the growth of the pension rolls and the rise in retirement costs.

Early retirement is, in part, a function of the danger and stress of policing and fire fighting. But several other incentives are at work. By far the most significant are the eligibility requirements. Many cities permit uniformed personnel to retire at half-pay after twenty or twenty-five years of service at any age. Others impose an age requirement, but it is usually forty-five or fifty and rarely more than fifty-five. In most cities, an officer can also retire on at least half-pay at any age if he has a service-connected disability. And in many cities the retirement board is required to presume that an officer who has heart or lung trouble sustained it in the line of duty.

Moreover, firemen and policemen have a strong incentive to retire in their middle or late forties because, as one actuarial firm put it, if they wait much longer, "the possibilities of other employment are significantly reduced."[1] Also reduced are the opportunities to pick up enough credit to qualify for Social Security. A full cost-of-living escalator is another incentive for early retirement. If retirement allowances are protected against inflation, many retired officers can get by without taking another job—especially if they are willing to move to the countryside, where the cost of living is lower.

Commercial and civic groups have long criticized early retirement. And although they have prevailed on the authorities to tighten eligibility requirements in some cities and to offer veteran officers financial incentives to stay on the job in others, firemen and policemen still retire roughly fifteen years earlier than nonuniformed employees. The persistence of early retirement is a result of more than just the political clout of the uniformed forces. It is also a consequence of the military analogy, the view of policing and fire fighting that emerged in the late nineteenth and early twentieth centuries and became the conventional wisdom of the upper middle and upper classes not long after. According to this view, policing and fire fighting are quasi-military activities that can be carried out only by young men in peak physical condition. The work is so hazardous and arduous that after twenty or twenty-five years most officers will be unable to do their part in the wars against crime and conflagration. The military analogy provided the rationale for treating firemen and policemen differently from nonuniformed employees and thus laid the foundation for the two-tiered pension system.

Over the years, the authorities have taken a few steps that have eased the firemen's and policemen's pension problem. They have put the pension systems on a contributory basis and raised the rates to the point where uniformed officers contribute about as large a share of their salaries as nonuniformed employees. In most cities, they have also replaced the pay-as-you-go approach with a reserve system, under which the government puts in enough money to pay long-term costs and sometimes to amortize the unfunded liabilities as well. In a few cities, the authorities have even imposed an age requirement for service pensions, capped the cost-of-living escalator, and forced new recruits into a statewide retirement plan for uniformed officers.

But in most cities these steps have not solved the pension problem, the crux of which is the immense growth of the pension rolls. To slow down their growth, the authorities will have to deter early retirement, which means that they will have to tighten eligibility requirements and, if need be, lower retirement benefits. This is no mean task, for it requires that the authorities abolish what remains of the two-tiered pension system, treat uniformed personnel in much the same way as nonuniformed employees, and above all disavow the view of policing and fire fighting on which current policy is based.

The authorities will probably find it very hard to tighten eligibility requirements and lower retirement benefits, especially if they try to impose these changes on current members. Although the political constraints on pension reform are weaker now than at any time in the recent past, the legal constraints are as strong as ever. These constraints are an outgrowth of the change in the legal status of public employee pensions that was brought about in the second third of the twentieth century by a series of constitutional amendments, state laws, and judicial decisions. Originally regarded as gratuities or bounties, pensions were redefined as contractual obligations. As such, they are protected by Article I of the Constitution, which prohibits the state from impairing these obligations, and by the Fourteenth Amendment, which prevents the state from taking property without due process of law.

As I observed earlier, there is something perplexing about the contract theory. It is not clear why the authorities should be allowed to raise benefits but not to lower them, to make mistakes but not to correct them. Nor is it clear why employees should be entitled not only to benefits promised at the time they entered the service but also to all improved benefits granted afterward. What is clear is that, while the contract theory protects the legitimate expectations of uniformed officers, it also prevents the authorities from doing very much about early retirement.

There have been several attempts to get around the contract theory, one of which is worth discussing. Late in the 1970s, Town Hall of California, an upper middle- and upper-class civic group, formed a committee to study the Los Angeles city and county pension systems, whose financial condition was the subject of much concern. The committee came up with several recommendations, among them a reduction in retirement benefits and/or an increase in contribution rates. Before submitting its final report, the committee asked O'Melveny and Myers, a prominent Los Angeles law firm, whether the proposed revisions could be imposed on current employees as well as future employees. O'Melveny and Myers pointed out that under California law a pension is a contractual obligation, a form of deferred compensation in which the employee has a vested right. Hence a reduction in benefits would be unconstitutional unless accompanied by a commensurate increase. But O'Melveny and Myers drew an ingenious distinction between benefits that have already been earned and benefits that would be earned in the future. And while admitting that California has no case law on the precise point, it held that the proposed revisions could probably be imposed on current members provided that they applied solely to the benefits that would be earned after the revisions went into effect.[2]

Following the advice of O'Melveny and Myers, reform groups drafted a charter amendment in 1982 that put a 3 percent ceiling on the cost-of-living escalator of the firemen's and policemen's pension systems. Known as Amendment H, it was aimed not at new recruits, whose escalator had been capped in 1980, but at veteran officers, whose escalator was still pegged to the consumer price index (CPI). Under Amendment H, firemen and policemen would get a full cost-of-living adjustment for the benefits that had already accrued but only a 3 percent adjustment for the benefits that would accrue later. The firemen and policemen strongly objected. Charging that the city was breaking its contract with the uniformed forces, they claimed that the amendment would destroy the morale of the fire and police departments, undermine public safety in the city, and in the long run save the taxpayers very little money. The reformers countered that firemen's and policemen's pension costs were out of control, that a full cost-of-living adjustment encouraged officers to leave the service in the prime of life, and that even with a 3 percent ceiling the Los Angeles pension system would be among the most liberal around. Unless the amendment passed, its sponsors insisted, the city would have to make sharp cuts in vital services. The firemen and policemen spent roughly $350,000 to defeat the amendment. But the voters, worried about the rising costs and mounting deficits of the pension system, endorsed it by a margin of more than two to one.[3]

The firemen and policemen have asked the courts to nullify the amendment on the grounds that they have vested rights in the full cost-of-living escalator. How the courts will decide is hard to tell. A superior court recently upheld a similar charter amendment that put a 2 percent cap on the cost-of-living adjustment of the Pasadena firemen's and policemen's pension systems.[4] But the officers took the case to the appellate court, which has not yet issued an opinion; and the case will probably end up in the California Supreme Court, which has long adhered to the contract theory. Whatever the verdict on Amendment H, it is vital for the courts to come up with a position that is not so restrictive as the contract theory or so anachronistic as the gratuity theory. The Los Angeles charter amendment provides one alternative. Another is offered by the Michigan Constitution—an amendment to which defines only accrued benefits as contractual obligations. Still another might be to guarantee employees only as much in benefits as they were promised at the time they entered the service. Whether the courts come up with a position that takes into account not only the legitimate expectations of public employees but also the financial condition of their pension plans remains to be seen. Until they do, most cities will be hard pressed to tighten eligibility requirements and lower retirement allowances in ways that will put an end to early retirement in the uniformed services.

Even if the courts weaken the legal constraints on pension reform, it is far from certain that the changes will solve the pension problem. The history of urban America is full of reforms that did not work out as anticipated or that had profound unintended consequences—and so is the history of its firemen's and policemen's pension plans. A few examples should suffice. The imposition of the merit system made it much harder to remove firemen and policemen for political reasons; while this insulated officers from partisan politics, it also lowered the turnover rate and thereby helped swell the pension rolls. Similarly, the imposition of maximum age requirements reduced the average entry age of firemen and policemen; although this may have improved the caliber of the rank and file, it also lowered the age at which officers retire and thus raised the number of years they spend on the pension rolls. In Los Angeles, the campaign to impose an age requirement for service pensions went awry and culminated in the adoption of a charter amendment that left eligibility requirements intact, raised retirement benefits, and created a cost-of-living escalator. In Washington, the attempts to consolidate local firemen's and policemen's pension systems produced an even more liberal statewide retirement plan for uniformed officers and instead of easing the pension burden simply shifted it from the cities to the state.

If the authorities decide to dismantle what remains of the two-tiered pension system and hereafter treat firemen and policemen in much the same way as nonuniformed employees, the results may include a few unintended consequences that are worth spelling out. One has already been discussed at length. In an attempt to slow down the growth of the pension rolls, reform groups have long pressed the authorities to impose an age requirement for service pensions. Such a requirement might lead some officers to delay retirement by five, ten, or even fifteen years, thus raising the average age at which they retire and lowering the number of years they spend on the pension rolls. But many others, reluctant to wait for a service pension, might instead apply for a disability pension. If so, an age requirement would be less likely to slow down the growth of the pension rolls than to drive up the number of disability pensions. Things might work out as anticipated if the authorities also tighten the eligibility requirements for disability pensions, abolish the heart and lung laws, define disability more narrowly, screen applications more closely, reduce employee representation on retirement boards, institute annual medical checkups for retirees, and push the uniformed forces to find work for officers who are less than 100 percent fit. But the prospects for these reforms are poor.

Moreover, if the authorities treat uniformed officers in the same way as nonuniformed employees when it comes to eligibility requirements and retirement benefits, why should they treat them differently when it comes to other things? Consider vesting. At the outset, firemen's and policemen's pension systems were nonvested: an officer could not receive a pension (or any part thereof) until he had fulfilled all the requirements. A nonvested plan was favored largely on the ground that it would discourage veteran officers from leaving the force in their most productive years. Some firemen's and policemen's pension systems subsequently adopted vesting. But overall these systems are less likely to be vested than retirement plans for nonuniformed employees. According to a nationwide survey of locally administered plans, only 62 percent of firemen and policemen, as opposed to 97 percent of nonuniformed employees and 100 percent of teachers, have vesting.[5] If vesting is a quid pro quo for tighter eligibility requirements and lower retirement benefits, many officers may leave the service after five, ten, or fifteen years. Hence vesting might swell the pension rolls and thereby exacerbate the pension problem.

Finally, if the authorities abolish the remnants of the two-tiered pension system, firemen and policemen might well join forces with nonuniformed personnel to push for more liberal retirement provisions for all public employees. This would be a striking change. Hitherto each group has looked out for its own interests. Firemen and policemen have often justified their demands for extremely liberal benefits on the ground that they have little in common with nonuniformed employees. For their part, nonuniformed employees have argued that their work is, in its own way, as demanding as policing and fire fighting and that they are entitled to the same benefits as uniformed personnel. On occasion, their interests have coincided. New York City's firemen, policemen, and nonuniformed employees worked together to win passage of a constitutional amendment that defined public employee pensions as contractual obligations. Milwaukee's public servants, uniformed and nonuniformed, mounted a long and successful campaign on behalf of a bill that gave them vested rights in their retirement systems. But these occasions are rare. Hence the abolition of the two-tiered pension system might help solve the firemen's and policemen's pension problem, but if it prompted uniformed and nonuniformed personnel to join forces, it might aggravate the general municipal employee pension problem, which would leave the cities in even worse shape.

Spokesmen for firemen and policemen have long criticized proposals to tighten eligibility requirements, lower retirement benefits, and otherwise discourage early retirement. If these proposals are adopted, they have argued, the uniformed services will be unable to attract qualified recruits, who will apply for less arduous and less dangerous positions in the public service or look for more remunerative jobs in the private sector. If eligibility requirements are tightened, forcing superior officers to remain in the service until their middle or late fifties, the departments will find it hard to promote able patrolmen and fire fighters, who will become demoralized by the lack of opportunity. And if the retirement benefits are lowered, the departments will find it hard to retain veteran officers, who will be tempted to leave the service after five or ten years if an attractive opportunity arises elsewhere. Most significant of all, these reforms will make it far harder for the fire and police departments to get rid of officers in their middle and late fifties, even though many of them may be superannuated and no longer able to carry their weight. These changes, spokesmen for the firemen and policemen have warned, will so lower the caliber of the uniformed forces that they will be unable to maintain public safety in urban America.

At the heart of this argument is the military analogy. If the uniformed services are engaged in a war against crime and conflagration, it follows that policing and fire fighting are "a young man's work," as Edward J. Kiernan, president of the International Conference of Police Associations, told the House Subcommittee on Social Security in 1976.[6] It also follows that after twenty or twenty-five years of service most uniformed officers will be pretty much worn out, unable to do the job, and that the fire and police departments must have some humane way to get rid of them and bring in younger men to take their place. Lastly, if the firemen's and policemen's jobs are as arduous and dangerous as the soldiers', it follows that the promise of a liberal pension is vital to the uniformed forces' efforts to attract qualified recruits, retain veteran officers in their prime, and encourage superannuated officers to leave the service.

But it is far from clear that the conventional wisdom is correct. There is no compelling evidence that policing and fire fighting are "a young man's work." A corollary of the military analogy, this position is simply an article of faith. It may be, as Kiernan claimed, that most persons "do not want a 60-year-old cop chasing a bank robber down the street [or] a 60-year-old fireman trying to climb a ladder to carry a woman out of a blazing building."[7] Even so, that is no reason to allow officers to retire on a service pension in their early or mid-fifties, much less in their mid- or late forties, which is the practice in most big cities and the principal source of the pension problem. Nor is there a reason to shelve proposals to tighten eligibility requirements or otherwise reduce disparities in the treatment of uniformed officers and nonuniformed personnel. Moreover, many policemen spend their time on much less arduous and dangerous tasks than chasing bank robbers. The fire departments also have jobs that do not include climbing ladders, although not nearly as many. There is no reason that many of these jobs cannot be done by officers in their late fifties and early sixties—or for that matter by younger officers whose disabilities prevent them from taking on more difficult and hazardous tasks.

Even if the conventional wisdom is correct, it is far from clear that big cities can afford to allow, much less to encourage, uniformed officers to retire so much earlier than nonuniformed employees. At a time when most cities are finding it very hard to make ends meet, it is extremely difficult to justify a policy that forces them to support two fire departments and two police departments. Unless the cities are prepared to defer capital improvements, put off regular maintenance, cut the municipal work force, and otherwise slash vital services, they have no choice but to hold down the growth of the pension rolls (as well as the increase in the retirement allowances). To do so, it will be necessary to tighten eligibility requirements and lower retirement benefits—and, by implication, to repudiate the military analogy, abolish the remnants of the two-tiered pension system, and treat firemen and policemen in much the same way as nonuniformed personnel. These reforms may reduce the efficiency of the uniformed forces by making it harder to attract able recruits, promote qualified officers, and get rid of superannuated veterans. But if this is a high price to pay for a solution to the pension problem, it is not a prohibitive one.

Notes

Introduction

1. *Los Angeles Times*, March 13, 1972; Los Angeles Board of Pension Commissioners, *Annual Report: 1971*, p. 28, *Annual Report: 1972*, pp. 5, 9, 28, 32, 39; Los Angeles City Controller, *Annual Report: 1972*, p. 40, schedule C-3; Los Angeles Department of Pensions, "Fire and Police Pension System" (1975), p. 8, Department of Pension Files, Los Angeles.

2. Los Angeles Board of Pension Commissioners, *Annual Report: 1977*, pp. 6–7, 10–11, 30, 42; Los Angeles City Controller, *Annual Report: 1977*, pp. 3, 9, schedule C-3; *Los Angeles Times*, November 3, 1977.

3. U.S. Bureau of the Census, *Census of Governments, 1977: Volume 6, Topical Studies: Number 1, Employee-Retirement Systems of State and Local Governments* (Washington, D.C., 1978), pp. 30–55; letter to the author from Paul Soullier, Detroit Auditor General's Office, December 20, 1979; letter to the author from Thomas J. O'Brien, Office of the City Administrator, Washington, D.C., November 21, 1979; Richard F. Camus and Associates, "Actuarial Study of the New Orleans Fire and Police Pension and Relief Funds as of June 30, 1977" (1978), pp. 40F–43F, 20P–25P; letter to the author from Richard W. Brune, New Orleans Department of Finance, March 11, 1980; Coates, Herfurth & England, "Actuarial Valuation of the San Francisco City and County Employees' Retirement System as of June 30, 1978" (1979), schedules 1 and 3; *Oakland Tribune*, January 15, 1976; interview with Jonathan Schwartz, New York City Employees' Retirement System, November 26, 1979. Data are for 1977 in all cities except Washington, D.C., and San Francisco, where they are for 1978.

4. U.S. Bureau of the Census, *Retirement Systems for State and Local Government Employees: 1941* (Washington, D.C., 1943), pp. 106–11; U.S. Bureau of the Census, *1972 Census of Governments: Volume 6, Topical Studies: Number 1, Employee Retirement Systems of State and Local Governments* (Washington, D.C., 1973), pp. 30–53; Bureau of the Census, *Employee-Retirement Systems of State and Local Governments: 1977*, pp. 30–55; Schwartz interview, November 26, 1979; Geo. E. Engel to Detroit City Council, memorandum dated January 13, 1942, pp. 11, 18, 22, 26, Detroit Auditor General's Files, Detroit; Soullier letter, December 20, 1979; *Washington Post*, February 26, 1978; U.S. Congress, Senate Committee on the District of Columbia, *Hearings [on] Fiscal Pressures on the District of Columbia: Part 3. Pension Systems* (Washington, D.C., 1976), pp. 288–92; U.S. Treasury Department, Office of Debt Analysis, "D.C. Metropolitan Policemen and Firemen Retirement Systems: Actuarial Valuation as of December 31, 1971," app. B, tables B6-TH, B6-TL.

5. U.S. Bureau of the Census, *1967 Census of Governments: Volume 6, Topical Studies: Number 2, Employee-Retirement Systems of State and Local Governments* (Washington, D.C., 1968), pp. 32–59; Bureau of the Census, *Employee-Retirement Systems of State and Local Governments: 1977*, pp. 30–55; Schwartz interview, November 26, 1979; Bureau of the Census, *Retirement Systems for State and Local Government Employees: 1941*, pp. 98, 106; Los Angeles Board of Pension Commissioners, *Annual Report: 1978*, p. 42; letter to the author from Wm. S. Hutchison, Los Angeles Fire and Police Pension System, December 21, 1979; O'Brien letter, November 21, 1979; Soullier letter, December 20, 1979; U.S. Bureau of the Census, *City Government Finances in 1976–77* (Washington, D.C., 1979), p. 38. Figures are for 1977 in all cities except for New York and Washington, D.C., where they are for 1978.

6. Bureau of the Census, *Employee-Retirement Systems of State and Local Governments: 1977*, pp. 30–55; Schwartz interview, November 26, 1979; Bureau of the Census, *City Government Finances in 1976–77*, pp. 14, 16, 47; Oakland Police and Fire Retirement Systems, *Twenty-Sixth Annual Report: 1977*, p. 3; Detroit Policemen and Firemen Retirement System, *Thirty-Seventh Annual Report: 1978*, p. 17; Soullier letter, December 20, 1979; *San Francisco Examiner*, November 29, 1979; San Francisco City and County Employees' Retirement System, *Annual Report: 1974–1975*, pp. 4, 9, *Annual Report: 1978–1979*, pp. 31, 46. New York City payroll data are for 1978.

7. Bureau of the Census, *Employee-Retirement Systems of State and Local Governments: 1977*, pp. 30–55; John D. MacPhail, "Actuarial Report on the City of Portland, Oregon Fire and Police Disability and Retirement Fund" (1976), p. 2; Milliman & Robertson, Inc., "Fire and Police Disability and Retirement Fund of the City of Portland, Oregon. Actuarial Valuation (as of July 1, 1979)" (1979), p. 3; telephone conversation with Robert P. Clohessy, Portland Fire and Police Disability and Retirement Fund, February 25, 1980; *New Orleans Times-Picayune*, May 17, 1974; O'Brien letter, November 21, 1979; Senate Committee on the District of Columbia, *Hearings on Fiscal Pressures on the District of Columbia*, pp. 301–3; U.S. Congress, Senate Committee on Governmental Affairs, Subcommittee on Governmental Efficiency and the District of Columbia, *Hearings [on the]*

District of Columbia Pension Reform Act (Washington, D.C., 1978), p. 693. Figures for New Orleans refer to the firemen's and policemen's pension systems that cover members appointed before 1968. The systems that cover members appointed in 1968 or after operate on a reserve basis.

8. Schwartz interview, November 26, 1979; Los Angeles Board of Pension Commissioners, *Annual Report: 1978*, p. 30; Senate Subcommittee on Governmental Efficiency and the District of Columbia, *Hearings on the District of Columbia Pension Reform Act*, p. 691; Detroit Policemen and Firemen Retirement System, *Thirty-Seventh Annual Report: 1978*, pp. 5–6; telephone conversation with Walter L. Johnson, Oakland Municipal Employees' Retirement System, December 18, 1980; *Portland Oregonian*, September 11, 1976; Richard F. Camus and Associates, "Actuarial Study of the New Orleans Fire and Police Pension Funds," pp. 13F–14F, 40F–41F, 11P–13P, 20P–23P; Brune letter, March 11, 1980; Legislative Commission on Pensions and Retirement, *Report to the 1979–1980 Minnesota State Legislature* (1980), pp. 60–64; U.S. Bureau of the Census, *Finances of Employee-Retirement Systems of State and Local Governments in 1975–76* (Washington, D.C., 1977), pp. 23, 30; U.S. Bureau of the Census, *1977 Census of Governments: Volume 2, Taxable Property Values and Assessment/Sales Price Ratios* (Washington, D.C., 1978), pp. 98, 115; U.S. Bureau of the Census, *Property Values Subject to Local General Property Taxation in the United States: 1978* (Washington, D.C., 1979), p. 40; *New York Daily News*, March 16, 1973. Data for unfunded liabilities are for 1976 in Portland and Washington, D.C.; for 1977 in Oakland, New Orleans, and New York; for 1978 in Detroit, Minneapolis, and Los Angeles. Figures for assessed value are for 1977 except in Minneapolis, where they are for 1978.

9. U.S. Congress, House Committee on Education and Labor, *Pension Task Force Report on Public Employee Retirement Systems* (Washington, D.C., 1978), pp. 51, 130, 165, 393.

10. *Ibid.*, pp. 289, 330–31; *Wall Street Journal*, May 5, 1978. The Pension Task Force's figures, which included officers who were covered by separate firemen's and/or policemen's pension systems, but not officers who belonged to consolidated municipal employee retirement plans, probably understated by about two-fifths the active and retired members, the pension payroll, and the employer contributions.

11. House Committee on Education and Labor, *Pension Task Force Report*, pp. 330–31; James F. Day, *Teacher Retirement in the United States* (North Quincy, Mass.: Christopher, 1971), pp. 39–41; Bureau of the Census, *Employee-Retirement Systems of State and Local Governments: 1977*, pp. 30–55.

12. *Los Angeles Times*, March 13, 1972; Los Angeles Board of Pension Commissioners, *Annual Report: 1977*, pp. 6, 7, 10, 12, 30, 34, 42; Los Angeles City Employees' Retirement System, *Annual Report: 1977*, pp. 1, 3–5, 8; Los Angeles City Controller, *Annual Report: 1977*, schedule C-3.

13. Bureau of the Census, *Employee-Retirement Systems of State and Local Governments: 1977*, pp. 30–55; Bureau of the Census, *Finances of Employee-Retirement Systems of State and Local Governments in 1975–76*, p. 28; Richard F. Camus and Associates, "Actuarial Study of the New Orleans Fire and Police Pension Funds,"

pp. 40F–43F, 20P–25P; Brune letter, March 11, 1980; Soullier letter, December 20, 1979; San Francisco City and County Employees' Retirement System, *Annual Report: 1977–1978*, p. 9; Minnesota State Planning Agency, Office of Local and Urban Affairs, *Minneapolis–St. Paul Study. Part II: Report on Pensions* (1978), p. 32.

14. U.S. Congress, House Committee on Ways and Means, *Hearings [on] General Revenue Sharing* (Washington, D.C., 1971), p. 290; Barbara A. Patocka, "Will Pension Costs Push America's Cities Over the Brink?" *Institutional Investor*, June 1975, pp. 55, 57; "The Hidden Costs that May Break the Cities," *Nation's Business*, September 1971, p. 31; Dean Lund, "Fire and Police Pension Funds: An Agonizing Reappraisal," *Minnesota Municipalities*, September 1971, p. 270; *Boston Globe*, October 17, 1976; *Wall Street Journal*, June 25, 1973.

15. Senate Committee on Governmental Affairs, *Hearings on the District of Columbia Pension Reform Act*, p. 13. See also Barbara A. Patocka, "The Herculean Task Is Underway," *Pensions*, May–June 1973, pp. 33–48.

16. House Committee on Education and Labor, *Pension Task Force Report*, pp. 7–13.

17. Advisory Commission on Intergovernmental Relations, *City Financial Emergencies: The Intergovernmental Dimension* (Washington, D.C., 1977), pp. 38–39; U.S. Congress, House Committee on the District of Columbia, Subcommittee on Incorporations, *Report of Hearings on H.R. 22322* (Washington, D.C., 1910), pp. 5–6, 35–36; *Congressional Record*, May 11, 1912, p. 6260, August 3, 1916, p. 12045; *Greater Cleveland*, October 21, 1943, pp. 27–28, October 3, 1946, p. 9; House Committee on Education and Labor, *Pension Task Force Report*, pp. 97–98.

18. *New Orleans Times-Picayune*, August 16, 1973; John C. Perham, "The Mess in Public Pensions," *Dun's Review*, March 1976, p. 48.

19. U.S. Congress, Joint Economic Committee, *The Current Fiscal Condition of Cities: A Survey of 67 of the 75 Largest Cities* (Washington, D.C., 1977), p. 2; David Grossman, *The Future of New York City's Capital Plant* (Washington, D.C.: Urban Institute, 1979), pp. 76–87, 91–92; David Berkowitz, "Neglect's Revenge: A Winter Water Tale," *Boston Globe*, February 18, 1979; Neal R. Pierce, "Years of Neglect, Bills Are Now Coming Due," *Boston Globe*, October 30, 1978; Nancy Humphrey, George E. Peterson, and Peter Wilson, *The Future of Cleveland's Capital Plant* (Washington, D.C.: Urban Institute, 1979), pts. 2 and 3; George E. Peterson, "Capital Spending and Capital Obsolescence: The Outlook for Cities," in Roy Bahl, ed., *The Fiscal Outlook for Cities* (Syracuse, N.Y.: Syracuse University Press, 1978), pp. 54–57, 62–67.

20. *New York Times*, March 31, 1977, December 28, 1978, September 2 and 17, November 9, 1979, October 13, 1980; *U.S. News & World Report*, October 27, 1975, pp. 15–16; Roy Bahl, Bernard Jump, Jr., and Larry Schroeder, "The Outlook for City Fiscal Performance in Declining Regions," in Bahl, ed., *Fiscal Outlook for Cities*, pp. 17–18; Richard P. Nathan et al., "Monitoring the Public Service Employment Program: The Second Round," report for the National Commission for Manpower Policy (1979), pp. 10–13.

21. Robert M. Fogelson, *Big-City Police* (Cambridge, Mass.: Harvard Univer-

sity Press, 1977), pp. 274–75; *New York Times*, February 27, 1978; Michael S. Serrill, "Urban Crisis Is Making Police Vulnerable—and Angry," *Police Magazine*, Summer 1977, pp. 2, 4, 8–11; telephone conversation with Sergeant Donald Nash, Public Information Unit, Detroit Police Department, December 18, 1980.

22. Fogelson, *Big-City Police*, pp. 274–75; Jack Stieber, *Public Employee Unionism: Structure, Growth, Policy* (Washington, D.C.: Brookings Institution, 1973), pp. 59–66; Richard T. Margolis, *Who Will Wear the Badge?* report of the U.S. Commission on Civil Rights (1971), pp. 2–3; *Boston Globe*, September 9 and 13, 1980.

23. Bureau of the Census, *Employee-Retirement Systems of State and Local Governments: 1977*, pp. 38, 46; Schwartz interview, November 26, 1979; Bureau of the Census, *City Government Finances in 1976–77*, p. 47; San Francisco City and County Employees' Retirement System, *Annual Report: 1978–1979*, p. 46; Coates, Herfurth & England, "Actuarial Valuation of the San Francisco Employees' Retirement System as of June 30, 1978," pp. 54–61; Bureau of the Census, *City Government Finances in 1977–78*, p. 19; U.S. Bureau of the Census, *Finances of Employee-Retirement Systems of State and Local Governments in 1977–78* (Washington, D.C., 1979), pp. 16, 20, 33, 34; Richard F. Camus and Associates, "Actuarial Study of the New Orleans Fire and Police Pension Funds," pp. 40F–41F, 19P–23P; Brune letter, March 11, 1980; Detroit Policemen and Firemen Retirement System, *Thirty-Seventh Annual Report: 1978*, p. 17; Soullier letter, December 20, 1979. Figures for New York and New Orleans are for 1977; for San Francisco and Detroit they are for 1978.

1. The Two-Tiered Pension System

1. Roger Lane, *Policing the City: Boston 1822–1885* (Cambridge, Mass.: Harvard University Press: 1967), pp. 182–83; William Graebner, *A History of Retirement: The Meaning and Function of an American Institution, 1885–1978* (New Haven, Conn.: Yale University Press, 1980), pp. 9–13; David Hackett Fischer, *Growing Old in America* (New York: Oxford University Press, 1977), pp. 167–71; W. Andrew Achenbaum, *Old Age in the New Land* (Baltimore: Johns Hopkins University Press, 1978), pp. 48–49; U.S. Department of Labor, Bureau of Labor Statistics, *Public Service Retirement Systems* (Washington, D.C., 1929), pp. 176–233; William Henry Glasson, *Federal Military Pensions in the United States* (London, 1918), pt. 1, chs. 2–5.

2. *Congressional Record*, March 20, 1882, pp. 2047–51; March 21, 1882, pp. 2100–3; April 27, 1882, pp. 3365–75; June 19, 1882, pp. 5100–5; 47th Cong., 1st sess., app., p. 429.

3. Richard B. Calhoun, "From Community to Metropolis: Fire Protection in New York City 1790–1875" (Ph.D. diss., Columbia University, 1973), pp. 152–53; Clarence H. Forrest, *Official History of the Fire Department of the City of Baltimore* (Baltimore, 1898), ch. 10; Joseph M. Crowley, "Report to Special Committee on Police and Fire Pensions Relative to the Laws Affecting the Same," *Cleveland City Record*, June 9, 1937, app., p. 1230; *Firemen's Journal*, April 5,

1879, p. 282; George Austin Ketcham, "Municipal Police Administration: A Comparative Study of Law Enforcement in Cincinnati, Chicago, New Orleans, New York and St. Louis, 1844–1877" (Ph.D. diss., University of Missouri, 1967), p. 218; Henry Mann, ed., *Our Police: A History of the Providence Police from the First Watchman to the Last Appointee* (Providence, 1889), ch. 21; Richard Sylvester, comp., *District of Columbia Police* (Washington, D.C., 1894), ch. 12; *Kansas City Star*, January 22, 1895, June 1, 1897; *Firemen's Standard*, January 1881, p. 2, February 1884, p. 4; Paul Studensky, *Teachers' Pension Systems in the United States* (New York, 1920), pp. 4–15.

4. Robert M. Fogelson, *Big-City Police* (Cambridge, Mass.: Harvard University Press, 1977), ch. 1.

5. *Ibid.;* Calhoun, "From Community to Metropolis," chs. 2–4; Andrew H. Neilly, "The Violent Volunteers: A History of the Volunteer Fire Department of Philadelphia, 1736–1781" (Ph.D. diss., University of Pennsylvania, 1960), chs. 1–7; Arlen Ross Dykstra, "A History of St. Louis Firefighting: The Transitional Years from Volunteer to Professional Pattern 1850–1880" (Ph.D. diss., St. Louis University, 1970), ch. 1.

6. Calhoun, "From Community to Metropolis," ch. 7; Neilly, "The Violent Volunteers," ch. 13; Dykstra, "History of St. Louis Firefighting," ch. 2; *Firemen's Journal*, July 12, 1879, p. 29, July 3, 1880, pp. 21–22, and November 27, 1880, pp. 499–500; *Firemen's Standard*, February 16, 1894, p. 5, July 1, 1895, p. 14; *New Orleans Daily Picayune*, June 28, July 11, 1891.

7. *Firemen's Standard*, October 1879, p. 4; Crowley, "Report to Special Committee on Police and Fire Pensions," p. 1230; Herbert Asbury, *Ye Olde Fire Laddies* (New York: Knopf, 1930), p. 94; Lane, *Policing the City*, p. 183; Metropolitan Police Department of the District of Columbia, *Report of Operations for the Fiscal Year 1886* (Washington, D.C., 1886), pp. 31–32; Alexander R. Piper, *Report of an Investigation of the Discipline and Administration of the Police Department of the City of Chicago* (Chicago, 1904), p. 11.

8. *Firemen's Journal*, December 15, 1877, p. 74, April 23, 1881, p. 326; Howard O. Sprogle, *The Philadelphia Police, Past and Present* (Philadelphia, 1887), p. 241; *Proceedings of the Eighth Convention [of the] Massachusetts State Firemen's Association* (1887), pp. 37–38; *Rochester Democrat and Chronicle*, February 12, 1887; Fogelson, *Big-City Police*, ch. 2.

9. U.S. Congress, House Committee on the District of Columbia, *Report [on] Policemen and Firemen, District of Columbia*, 48th Cong., 1st sess., H.R. 979.

10. *Congressional Record*, June 11, 1894, pp. 6122–30.

11. Boston City Council, *Reports of Proceedings*, July 18, 1892, pp. 703–8, August 1, 1892, pp. 721–28.

12. *Firemen's Standard*, December 1879, p. 4; Boston City Council, *Reports of Proceedings*, July 18, 1892, p. 704, August 1, 1892, p. 724; *Rochester Democrat and Chronicle*, March 30, 1887; *Detroit Evening News*, May 29, 1885; Mary Wilcox Brown, *The Development of Thrift* (New York, 1900), pp. 146, 208; James Henry Hamilton, *Savings and Savings Institutions* (New York, 1902), pp. 27–28.

13. *St. Louis Post-Dispatch*, February 12 and 16, 1893.

14. Boston City Council, *Reports of Proceedings*, March 15, 1877, p. 159, July

18, 1892, p. 706, and August 1, 1892, p. 724; House Committee on the District of Columbia, *Report on Policemen and Firemen*.

15. *Rochester Democrat and Chronicle*, February 13, March 31, 1887; Rochester Bureau of Municipal Research, *Municipal Research*, March 1944, p. 415; Boston City Council, *Reports of Proceedings*, July 18, 1892, p. 706, August 1, 1892, pp. 721–25; *Detroit Evening News*, May 29, 1885; *Congressional Record*, June 11, 1894, pp. 6122–30.

16. *Firemen's Journal*, December 27, 1879, pp. 541–42; *St. Louis Post-Dispatch*, February 26, 1893; Providence City Council, *Report of the Joint Special Committee Relative to a Pension Fund for Disabled Policemen and Firemen with Accompanying Resolution* (Providence, 1893), pp. 4–6, 12–13; *Rochester Democrat and Chronicle*, March 31, 1887.

17. *Report of Major and Superintendent of [Washington, D.C.] Police* (1884), p. 12; Providence City Council, *Report of the Joint Special Committee Relative to a Pension Fund for Disabled Policemen and Firemen*, pp. 5–6; Boston City Council, *Reports of Proceedings*, March 15, 1877, p. 158, August 1, 1892, pp. 725–26; *St. Louis Post-Dispatch*, February 26, 1893.

18. Providence City Council, *Report of the Joint Special Committee Relative to a Pension Fund for Disabled Policemen and Firemen*, p. 6; U.S. Congress, House Committee on the District of Columbia, Subcommittee on Incorporations, *Report of Hearings on H.R. 22322* (Washington, D.C., 1910), p. 48; Boston City Council, *Reports of Proceedings*, March 15, 1877, p. 158; *St. Louis Post-Dispatch*, February 26, 1893; *Firemen's Journal*, November 29, 1884, pp. 434–36.

19. *St. Louis Post-Dispatch*, February 26, 1893; *Firemen's Journal*, April 23, 1881, p. 326; House Committee on the District of Columbia, *Report of Hearings on H.R. 22322*, pp. 54–56; Providence City Council, *Report of the Joint Special Committee Relative to a Pension Fund for Disabled Policemen and Firemen*, pp. 4–6.

20. *St. Louis Post-Dispatch*, February 26 and 27, 1893; *Firemen's Journal*, December 27, 1879, pp. 541–42; William McAdoo, *Guarding a Great City* (New York, 1906), p. 64; *Chicago Tribune*, February 3, 1887; House Committee on the District of Columbia, *Report of Hearings on H.R. 22322*, pp. 46–47.

21. Roy Lubove, *The Struggle for Social Security 1900–1935* (Cambridge, Mass.: Harvard University Press, 1968), pp. 12–13; Achenbaum, *Old Age in the New Land*, p. 49.

22. *Firemen's Journal*, June 13, 1885, p. 66.

23. Achenbaum, *Old Age in the New Land*, pp. 39, 45, ch. 3; Fischer, *Growing Old in America*, chs. 1–3; Graebner, *A History of Retirement*, p. 16; Joseph A. Lapp, "The Insurance of Thrift," in *The New American Thrift*, vol. 77 of *The Annals of the American Academy of Political and Social Sciences*, January 1920, pp. 21–26; Lubove, *Struggle for Social Security*, ch. 6.

24. *New York Times*, April 22, 23, and 27, 1879; New York State Senate Committee Appointed to Investigate the Police Department of the City of New York, *Report and Proceedings* (Albany, 1895), vol. 5, pp. 4737–43; *Chicago Tribune*, November 15, 1898; Illinois Senate Committee of Investigation, *Report on the Chicago Police System* (Springfield, Ill., 1898), p. 11.

25. Fogelson, *Big-City Police*, p. 59; McAdoo, *Guarding a Great City*, pp. 63–64.

26. Lewis Meriam, *Principles Governing the Retirement of Public Employees* (New York, 1918), pp. 3–17.

27. *Chicago Tribune*, February 3, 1887; *St. Louis Post-Dispatch*, February 26, 1893; *Firemen's Journal*, June 13, 1885, p. 66; *Firemen's Standard*, February 1, 1897, p. 12; Boston City Council, *Reports of Proceedings*, June 18, 1892, pp. 703–08; *Rochester Democrat and Chronicle*, February 12, 1887; Providence City Council, *Report of the Joint Special Committee Relative to a Pension Fund for Disabled Policemen and Firemen*, pp. 19–20; *Oregon Daily Journal*, May 27, 1913.

28. New York City Commission on Pensions, *Report on the Pension Funds of the City of New York* (New York, 1916), pt. 1, pp. 1–2; Lee Welling Squier, *Old Age Dependency in the United States* (New York, 1912), pp. 193–94.

29. Studensky, *Teachers' Pension Systems*, pp. 15–29; Commission on Pensions, *Report on the Pension Funds of the City of New York*, pt. 1, pp. 2–3; Boston City Council, *Reports of Proceedings*, January 25, 1892, pp. 72–73.

30. Massachusetts Commission on Old Age Pensions, Annuities and Insurance, *Hearings*, April 8, 1908, pp. 1, 4–5, 10, 15–18, 22–23, State House Library, Boston.

31. *Ibid.*, November 15, 1908, pp. 20–22, March 13, 1909, pp. 7–9; *Revised Record of the Constitutional Convention of the State of New York*, vol. 1 (Albany, 1900), pp. 857–62, 865–66, 978–83, 985, 990–91; *Debates of the Massachusetts Constitutional Convention 1917–1918*, vol. 3 (Boston, 1920), pp. 476–88, 491–92, 498–99, 526–27. See also Frederick L. Hoffman, "The Problems of Poverty and Pensions in Old Age," *American Journal of Sociology*, September 1908, pp. 193–94.

32. Massachusetts Commission on Old Age Pensions, *Hearings*, February 26, 1908, pp. 2–3, April 8, 1908, pp. 2–3, 13–14, 16–17, April 30, 1909, pp. 15–16, 33–36, and September 30, 1909, pp. 27–29; *Revised Record of the Constitutional Convention of the State of New York*, vol. 1, pp. 857, 993–94; *Debates of the Massachusetts Constitutional Convention*, vol. 3, p. 502.

33. Studensky, *Teachers' Pension Systems*, pp. 15–29; Squier, *Old Age Dependency*, p. 224; Commission on Pensions, *Report on the Pension Funds of the City of New York*, pt. 1, pp. 2–3; F. Spencer Baldwin, "Retirement Systems for Municipal Employees," in *Risks in Modern Industry*, vol. 38 of *The Annals of the American Academy of Political and Social Sciences*, July 1911, p. 7.

34. Baldwin, "Retirement Systems for Municipal Employees," pp. 6, 12; Edward F. Mason, "Municipal Pensions," *National Municipal Review*, April 1913, p. 264; Achenbaum, *Old Age in the New Land*, ch. 3; *Report of the [Massachusetts] Commission on Old Age Pensions, Annuities and Insurance* (Boston, 1910), p. 270; *Report of the Pennsylvania Commission on Old Age Pensions* (Harrisburg, 1919), pp. 179–80.

35. Baldwin, "Retirement Systems for Municipal Employees," p. 13; Meriam, *Principles Governing the Retirement of Public Employees*, pp. 3–17; Amos W. Butler, "Government and Municipal Pensions," in *Proceedings of the National Conference on Charities and Corrections* (1906), p. 485.

36. New York Bureau of Municipal Research, *A Report on the Police Pension Fund of the City of New York Submitted to the Aldermanic Committee on Police Investigation* (New York, 1913), pp. 17–18; Commission on Pensions, *Report on the Pension Funds of the City of New York*, pt. 1, p. 50, pt. 2, pp. 62–63, 90–91; *Report of the Illinois Pension Laws Commission* (Springfield, 1917), pp. 11–16, 272.

37. Baldwin, "Retirement Systems for Municipal Employees," pp. 8–9; Meriam, *Principles Governing the Retirement of Public Employees*, chs. 3, 5, 14; Paul Studensky, *The Pension Problem and the Philosophy of Contributions* (New York, 1917), pp. 3–18; George B. Buck, "Municipal Pensions and Pension Funds," in National Association of Comptrollers and Accounting Officers, *Proceedings of the Tenth Annual Convention* (1915), pp. 105–10.

38. Milton Conover, "Pensions for Public Employees," *American Political Science Review*, August 1921, pp. 351–55.

39. Baltimore Retirement Commission, *Report on a Proposed Retirement Plan for the Employees of the City of Baltimore* (Baltimore, 1925), p. 7; Baltimore Commission on Efficiency and Economy, *Report on Establishment of a Retirement System for the Employees of Baltimore City* (Baltimore, 1923), pp. 8–9; Oakland Civil Service Board, *A Proposed Retirement System for Municipal Employees of the City of Oakland, California* (Oakland, 1926), pp. 3, 5, 6.

40. *Report of the Pennsylvania Commission on Old Age Pensions*, p. 14; Conover, "Pensions for Public Employees," p. 352; George B. Buck, "Baltimore's New Retirement System," *National Municipal Review*, August 1926, p. 454; *[San Francisco] Municipal Record*, December 15, 1921, pp. 401–3; *San Francisco Journal*, February 5, 1922. See also Paul Studensky, "Pensions in Public Employment," *National Municipal Review*, April 1922, pp. 115–20.

41. Boston City Council, *Reports of Proceedings*, October 30, 1911, p. 427; Kenneth L. Webb, "Our Pension Bill," *Seattle Civil Service Journal*, November 1920, pp. 4, 8, 9; Ovid B. Blix, "Pensions for Public Employees," *Milwaukee Government Service*, November 1, 1936, pp. 4, 12; George A. Terhune, "The History of Our Retirement Plan," *Los Angeles City Employee*, June 1937, pp. 8–9, 26–27.

42. Anthony Pratt to Los Angeles City Council, June 23, 1922, Los Angeles City Council Files, 1922, no. 2404, Municipal Records Center, Los Angeles; Carnegie Foundation for the Advancement of Teaching, *Sixteenth Annual Report* (1922), pp. 145, 149; Webb, "Our Pension Bill," p. 4; Terhune, "Our Retirement Plan," p. 9; *[Los Angeles] Municipal Employee*, August 1929, p. 12.

43. Boston City Council, *Reports of Proceedings*, December 12, 1910, pp. 385–86; *Milwaukee Journal*, June 5, 1921; *Portland City Club Bulletin*, May 9, 1930, pp. 1–4; Terhune, "Our Retirement Plan," p. 9; *Oakland Tribune*, March 26, 1937.

44. William C. Beyer, "Municipal Civil Service in the United States," in Carl Joachim Friedrich et al., *Problems of the American Public Service* (New York: McGraw-Hill, 1935), pp. 135–36; Bureau of Labor Statistics, *Public Service Retirement Systems*, pp. 16, 25, 56; *[Portland] Civil Service*, April 1927, p. 3; *Oregon Voter*, October 21, 1944, p. 11; Graebner, *History of Retirement*, chs. 3, 7; Lubove, *Struggle for Social Security*, ch. 6.

45. *Milwaukee Journal*, September 4, 1935; *Milwaukee Sentinel*, September 7, 1935; Blix, "Pensions for Public Employees," pp. 4, 12.

46. *Los Angeles City Employee*, June–July 1935, p. 6; Terhune, "Our Retirement Plan," p. 9.

47. *San Francisco Examiner*, October 21, 1934; *Oakland Tribune*, March 26, 1937; *Milwaukee Journal*, April 27, October 30, 1936; *Milwaukee Sentinel*, July 16, 1937; *Milwaukee Leader*, August 3, 1937; Terhune, "Our Retirement Plan," pp. 9, 26–27; Herman Kehrli, "Portland Pension Problems," report to the Portland City Council (1934), p. 25.

48. Chas. P. Neill, *Pension Funds for Municipal Employees and Railroad Pension Systems in the United States*, U.S. Congress, Senate Documents, 61st Cong., 2d sess., no. 427, pp. 36–85; Bureau of Labor Statistics, *Public Service Retirement Systems*, pp. 56–73.

49. Neill, *Pension Funds for Municipal Employees*, pp. 36–85; Bureau of Labor Statistics, *Public Service Retirement Systems*, pp. 56–73; Buck, "Baltimore's New Retirement System," pp. 454–56.

50. Neill, *Pension Funds for Municipal Employees*, pp. 36–85; Bureau of Labor Statistics, *Public Service Retirement Systems*, pp. 56–73.

51. *Ibid.*

52. Glasson, *Federal Military Pensions*, p. 99; William Addleman Canoe, *The History of the United States Army* (New York: Appleton-Century, 1942), p. 262; Thomas Wilhelm, *A Military Dictionary and Gazetteer* (Philadelphia, 1881), pp. 485–86.

53. Robert Tilove, *Public Employee Pension Funds* (New York: Columbia University Press, 1976), pp. 273–74.

2. *Swelling Rolls and Soaring Costs*

1. U.S. Census Office, *Report on the Social Statistics of Cities in the United States at the Eleventh Census: 1890* (Washington, D.C., 1895), pp. 118–32.

2. Robert M. Fogelson, *Big-City Police* (Cambridge, Mass.: Harvard University Press, 1977), p. 30; Kansas City [Mo.] Public Service Institute, "Kansas City Firemen's Pension Fund" (1930), pp. 28–32.

3. *Los Angeles Times*, June 27, 1889; James F. Richardson, *The New York Police* (New York: Oxford University Press, 1970), p. 174; Chas. P. Neill, *Pension Funds for Municipal Employees and Railroad Pension Systems in the United States*, U.S. Congress, Senate Documents, 61st Cong., 2d sess., no. 427, pp. 40, 80; Cleveland Board of Police Commissioners, *Eleventh Annual Report: 1882*, p. 27; St. Louis Board of Police Commissioners, Annual Reports, 1880–89; Baltimore Board of Police Commissioners, Annual Reports, 1900–9; Commonwealth of Massachusetts, *Report of the Commission on Pensions* (Boston, 1914), pp. 233–39; Baltimore Board of Fire Commissioners, *Seventeenth Annual Report: 1900*, pp. 62–81.

4. Commonwealth of Massachusetts, *Report of the Commission on Pensions*, pp. 233–46; New York City Commission on Pensions, *Report on the Pension Funds of*

the City of New York (New York, 1916), pt. 1, pp. 142, 147, 150. New York City also had in 1914 nearly 2,200 disability pensioners, who on average served for roughly twenty years and retired at about forty-seven. They put in seven to eight years less than the service pensioners and went on the rolls seven to eight years earlier. Baltimore Board of Police Commissioners, *Annual Report: 1898–1899*, pp. 50–51; Philadelphia Chamber of Commerce, "Report on the Police Pension Fund" (1927), table D.

5. U.S. Public Health Service, *Vital Statistics of the United States, 1973. Volume II-Section 5. Life Tables* (Rockville, Md., 1975), p. 14; Commission on Pensions, *Report on the Pension Funds of the City of New York*, pt. 2, pp. 58, 84; J. D. Craig, *Report on the Firemen's Relief and Pension Fund, the Police Pension Fund and the Public School Teachers' Retirement Fund of the City of Buffalo* (1917), pp. 19, 38–39.

6. Commission on Pensions, *Report on the Pension Funds of the City of New York*, pt. 1, pp. 165–66; *Report of the Illinois Pension Laws Commission* (Springfield, 1917), pp. 11, 13; Craig, *Report of the Pension Funds of the City of Buffalo*, pp. 30–31, 46–47; Commonwealth of Massachusetts, *Report of the Commission on Pensions*, p. 27; U.S. Congress, Senate Documents, *Report of the Findings and Recommendations of the Committee on Retirement Policy for Federal Personnel*, 83d Cong., 2d sess., no. 89, pp. 193–94; U.S. Department of Labor, Bureau of Labor Statistics, *Public Service Retirement Systems* (Washington, D.C., 1929), pp. 151, 170.

7. Commission on Pensions, *Report on the Pension Funds of the City of New York*, pt. 1, pp. 165–66; *Report of the Illinois Pension Laws Commission*, pp. 11, 13; *Greater Cleveland*, April 11, 1928, p. 134; U.S. Bureau of the Census, *Financial Statistics of Cities Having a Population of Over 30,000: 1915* (Washington, D.C., 1916), pp. 92–93, 188–93, 224–25.

8. Metropolitan Police Department for the District of Columbia, *Report on Operations for the Fiscal Year 1886*, pp. 10–13; William F. Russell to John S. Clark, December 26, 1929, Chicago City Council Documents, vol. 1929, no. 308073, Chicago City Council Files, Chicago; Citizens' Police Committee, *Chicago Police Problems* (Chicago, 1931), pp. 249–69; James H. Hepbron, "Does Baltimore Need More Policemen?" study by the Baltimore Criminal Justice Commission (1940), pp. 1–13.

9. Citizens' Police Committee, *Chicago Police Problems*, pp. xiii–xiv; Chicago City Council, *Journal of the Proceedings*, January 23, 1918, pp. 1911–12; Chicago Bureau of Public Efficiency, "Shall the City of Chicago Employ Permanently 1,000 Additional Policemen?" (1919), pp. 1–12; Fogelson, *Big-City Police*, pp. 82, 119–20, 231, 274–75; International City Management Association, *Municipal Fire Administration* (Washington, D.C., 1968), pp. 158–67; Newark Bureau of Municipal Research, *Police Problems in Newark* (Newark, 1943), p. 19; George H. Taylor, ed., *Fire Protection Handbook* (Boston: National Fire Protection Association, 1962), ch. 10, p. 8.

10. Census Office, *Social Statistics of Cities: 1890*, pp. 118–32; *Monthly Labor Review*, December 1929, pp. 124–34, January 1930, pp. 118–26; International City Management Association, *The Municipal Year Book 1971* (Washington, D.C.,

1971), pp. 202–3, *The Municipal Year Book 1972* (Washington, D.C., 1972), p. 235. The twenty largest cities are the twenty largest cities in 1930, not in 1890 or 1970. In the case of the fire departments in 1890, I have excluded the six big cities that had a large number of call officers.

11. Census Office, *Social Statistics of Cities: 1890*, pp. 118–32; International City Management Association, *The Municipal Year Book 1971*, pp. 202–3, *The Municipal Year Book 1972*, p. 235.

12. National Association of Fire Engineers, *Twenty-Second Annual Convention* (1894), p. 142; Massachusetts State Firemen's Association, *Proceedings of the First Annual Convention* (1881), p. 6; Fogelson, *Big-City Police*, pp. 42–43, 49–51, 59–60.

13. Martin J. Schiesl, *The Politics of Efficiency: Municipal Administration and Reform in America, 1880–1920* (Berkeley: University of California, 1977), pp. 33, 38; Fogelson, *Big-City Police*, pp. 1–11, 61–66; Massachusetts State Firemen's Association, *Proceedings of the Eighth Annual Convention* (1887), p. 37; *Chicago Tribune*, March 29, April 2, 1895.

14. Fogelson, *Big-City Police*, pp. 79–80, 104–05, 184, 255; Schiesl, *Politics of Efficiency*, pp. 33–37, 44; U.S. Bureau of the Census, *General Statistics of Cities: 1915* (Washington, D.C., 1916), p. 96; U.S. Bureau of the Census, *Statistics of Fire Departments of Cities Having a Population of Over 30,000: 1917* (Washington, D.C., 1918), p. 69; Portland Police Survey Commission, "Survey of the Portland Police Bureau" (1934), p. 80; President's Commission on Law Enforcement and Administration of Justice, *Task Force Report: The Police* (Washington, D.C., 1967), p. 133.

15. District of Columbia Firemen's Association, Local 36, I.A.F.F., "Statement in Support of Improved Retirement Benefits," April 29, 1957, Records of the Government of the District of Columbia, Record Group 351, File 511b, National Archives, Washington, D.C. Cincinnati figures come from the annual reports of the Division of Police, Department of Safety, 1935–60; Los Angeles figures come from the annual reports of the Board of Pension Commissioners, 1950–70; Detroit figures come from the annual reports of the Police Department, 1930–50. Oakland Police and Fire Retirement System, *Twenty-First Annual Report*, p. 23; telephone conversation with Walter L. Johnson, Oakland Municipal Employees' Retirement System, November 27, 1979; interview with Jonathan Schwartz, New York City Employees' Retirement System, November 26, 1979.

16. New York Bureau of Municipal Research, *Report on a Survey of the Government of the City and County of San Francisco* (San Francisco, 1916), p. 172; Raymond B. Fosdick et al., *Criminal Justice in Cleveland* (Cleveland: Cleveland Foundation, 1922), pp. 26–28; August Vollmer, "Police Conditions in the United States," in National Commission on Law Observance and Enforcement, *Report on Police* (Washington, D.C., 1931), pp. 62–63; Citizens' Police Committee, *Chicago Police Problems*, pp. 51–52; National Association of Chiefs of Police of the United States and Canada, *Fourth Annual Convention* (1897), p. 31; Leonard V. Harrison, *Police Administration in Boston* (Cambridge, Mass.: Harvard University Press, 1934), p. 35; Public Administration Service, *The Selection of Firefighters* (Chicago, 1940), pp. 3–4; International City Managers' Association, *Municipal*

Fire Administration (Chicago, 1942), pp. 187–88; *Municipal Police Administration* (Chicago, 1943), pp. 118–19.

17. Boston City Council, *Reports of Proceedings*, January 18, 1875, pp. 66–67; Fogelson, *Big-City Police*, pp. 54–56; Alexander R. Piper, *Report of an Investigation of the Discipline and Administration of the Police Department of the City of Chicago* (Chicago, 1904), p. 11; *Philadelphia Inquirer*, June 23, 1924; William G. Shepard, "Crime in the Home of Its Friends," *Collier's*, December 5, 1925, p. 18.

18. Massachusetts Civil Service Commission, *First Annual Report* (1885), p. 27; Bureau of the Census, *General Statistics of Cities: 1915*, p. 96; Bureau of the Census, *Statistics of Fire Departments: 1917*, p. 68; Portland Police Survey Commission, "Survey of the Portland Police Bureau," pp. 41–42; International Association of Chiefs of Police, *Police Personnel Selection Survey* (Washington, D.C., 1968).

19. For Boston, see the annual reports of the Massachusetts Civil Service Commission, 1885–1915; for Cleveland, the annual reports of the Department of Public Safety, 1905–10; for Baltimore, the annual reports of the Board of Police Commissioners, 1946–55; for Milwaukee, the letter from Warren Fritschler to the author, May 8, 1978; for Oakland, the Oakland Police and Fire Retirement System, *Twenty-First Annual Report*, p. 23; president of Board of Commissioners of the District of Columbia to Harry B. Mitchell, February 4, 1943, Records of the Government of the District of Columbia, File 4-534.

20. Bureau of Municipal Research, *Report on the Government of San Francisco*, p. 172; Fosdick et al., *Criminal Justice in Cleveland*, p. 28; International City Managers' Association, *Municipal Police Administration*, p. 119.

21. Neill, *Pension Funds for Municipal Employees*, pp. 36–85; Val C. Mogensen to Members of the Police Pension Fund Committee, memorandum dated June 10, 1955, pp. 9–10, New Orleans Bureau of Governmental Research Files, New Orleans; Los Angeles Fire and Police Protective League, *History of the Department of Pensions of the City of Los Angeles* (1939), p. 1; Bureau of Labor Statistics, *Public Service Retirement Systems*, pp. 147–49, 166–68.

22. Neill, *Retirement System for Municipal Employees*, pp. 36–85; Commission on Pensions, *Report on the Pension Funds of the City of New York*, pt. 1, p. 2; Bureau of Labor Statistics, *Public Service Retirement Systems*, pp. 147–49, 166–68; Commonwealth of Massachusetts, *Report Submitted by the Legislative Research Council Relative to the Presumption Relative to Heart Disease and Retirement* (Boston, 1958), pp. 11, 14, 26–28; U.S. Congress, House Subcommittee on Fiscal Affairs and Committee on the District of Columbia, *Hearings and Markups [on] Financing Retirement Funds for Police, Firemen, Teachers, and Judges* (Washington, D.C., 1976), pp. 153–54; *Washington Post*, February 25, 1978; *Washington Star*, March 1, 1978.

23. "Service Pensions—Fire and Police," Los Angeles City Council Files, no. 21072, Municipal Records Center, Los Angeles; untitled memorandum dated April 23, 1954, New Orleans Bureau of Governmental Research Files; New York City Police Department, Annual Reports, 1943–52; Public Administration Service, "Compensation and Related Practices in Major United States Cities for Police and Fire Service" (1963), p. 56; U.S. Congress, Senate Committee on the District of

Columbia, *Hearings [on] Fiscal Pressures on the District of Columbia: Part 3. Pension Systems* (Washington, D.C., 1976), pp. 289–92.

24. Untitled actuarial report on the Los Angeles Fire and Police Pension System (1931), p. 44, Department of Pensions Files; George B. Buck, "Report on an Actuarial Investigation of the Police Relief Fund of the City of Cincinnati" (1930), also known as Report no. 24 of the Cincinnati Bureau of Governmental Research, pp. 25–26, and "Report on an Actuarial Investigation of the Firemen's Pension Fund of the City of Cincinnati" (1930), also known as Report no. 25 of the Cincinnati Bureau of Governmental Research, pp. 27–28; New York State Commission on Pensions, "Report of an Actuarial Investigation of the N.Y. Fire Department Relief Fund as of December 31, 1933" (1934), p. 34, and "Report of an Actuarial Investigation of the Police Pension Fund of the City of New York as of December 31, 1933" (1934), p. 34; U.S. Congress, House Committee on the District of Columbia, Subcommittee no. 3, *Hearings [on] Amendments to the Policemen's and Firemen's Retirement and Disability Act* (Washington, D.C., 1959), p. 53; U.S. Treasury Department, Office of Debt Analysis, "D.C. Metropolitan Policemen and Firemen Retirement Systems: Actuarial Valuation as of December 31, 1971," app. A, tables A12-P, A12-F; Coates, Herfurth & England, "Actuarial Investigation [of the San Francisco City and County Employees' Retirement System] as of June 30, 1976" (1978), p. 3.

25. U.S. Public Health Service, *Life Tables*, p. 14; Fogelson, *Big-City Police*, p. 26.

26. "Service Pensions—Fire and Police," Los Angeles City Council Files, no. 21072; untitled memorandum dated April 23, 1954, New Orleans Bureau of Governmental Research Files; Public Administration Service, "Compensation and Related Practices," p. 56; Senate Committee on the District of Columbia, *Hearings on Fiscal Pressures on the District of Columbia*, pp. 289–92.

27. Oakland Police and Fire Retirement System, *Twenty-Fourth Annual Report*, pp. 27, 29; telephone conversation with Walter L. Johnson, October 29, 1979; U.S. Public Health Service, *Life Tables*, p. 14.

28. Craig, *Report on the Pension Funds of the City of Buffalo*, pp. 30–31, 46–47; *Report of the Illinois Pension Laws Commission*, pp. 11, 13; U.S. Bureau of the Census, *Retirement Systems for State and Local Government Employees: 1941* (Washington, D.C., 1943), pp. 106–11; U.S. Bureau of the Census, *1972 Census of Governments: Volume 6, Topical Studies: Number 1, Employee Retirement Systems of State and Local Governments* (Washington, D.C., 1973), pp. 30–53; U.S. Bureau of the Census, *Census of Governments, 1977: Vol. 6, Topical Studies: Employee-Retirement Systems of State and Local Governments* (Washington, D.C., 1978), pp. 30–55.

29. See note 28.

30. Bureau of the Census, *Employee Retirement Systems of State and Local Governments: 1972*, pp. 30–53; Bureau of the Census, *Employee-Retirement Systems of State and Local Governments: 1977*, pp. 30–55; Schwartz interview, November 26, 1979; Conrad M. Siegel, Inc., "Act 293 Report. Municipal Pension Funds. 1976 Actuarial Study Analysis," report prepared for the Pennsylvania Department of Community Affairs (1977), schedules 1PCI, 1FCI; letter to the author

from Paul Soullier, Detroit Auditor General's Office, December 20, 1979; letter to the author from Thomas J. O'Brien, Office of the City Administrator, Washington, D.C., November 21, 1979; Coates, Herfurth & England, "Actuarial Valuation of the San Francisco City and County Employees' Retirement System as of June 30, 1978" (1979), schedules 1, 3; Richard F. Camus and Associates, "Actuarial Study of the New Orleans Fire and Police Pension and Relief Funds as of June 30, 1977" (1978), pp. 40F–43F, 20P–25P. Data are for 1977 for all cities except Los Angeles, Washington, D.C., New York, and San Francisco, where they are for 1978.

31. Neill, *Pension Funds for Municipal Employees*, pp. 36–85.

32. Harold L. Henderson, "Milwaukee Adopts Scientific Pension System for Police Department," *American City*, September 1922, p. 263; *Milwaukee Leader*, December 26, 1923; Los Angeles Fire and Police Protective League, *History of the Department of Pensions*, pp. 1–2; *Oakland Tribune*, April 8, 1951; *Los Angeles Examiner*, November 4, 1922; *Congressional Record*, May 22, 1957, pp. 7382–87, June 3, 1957, pp. 8197–210, August 10, 1957, pp. 14324–39.

33. U.S. Congress, Senate Committee on Governmental Operations, Subcommittee on Governmental Efficiency and the District of Columbia, *Hearings [on the] District of Columbia Pension Reform Act* (Washington, D.C., 1978) pp. 44–57, 64–84; District of Columbia Office of Personnel, Compensation and Research Division, "Police and Fire Retirement Practices," pp. 30–38, District of Columbia Office of Personnel Files, Washington, D.C.

34. Bureau of the Census, *General Statistics of Cities: 1915*, p. 69; Bureau of the Census, *Statistics of Fire Departments: 1917*, p. 51; George A. Ketcham, "Municipal Police Reform: A Comparative Study of Law Enforcement in Cincinnati, Chicago, New Orleans, New York and St. Louis, 1844–1877" (Ph.D. diss., University of Missouri, 1967), pp. 215–16; *Report of the [San Francisco] Board of Fire Commissioners: 1871*, p. 152, *1900*, p. 402; *Report of the City Auditor to the City Council of the City of St. Louis: 1874*, pp. 111–15; *Report of the [Boston City Council] Committee on Salaries: 1876*, p. 14; Boston Fire Department, *Annual Report: 1901*, p. 5.

35. Fogelson, *Big-City Police*, pp. 59, 82, 194; Bureau of the Census, *General Statistics of Cities: 1915*, p. 69; Bureau of the Census, *Statistics of Fire Departments: 1917*, pp. 51–53; *Monthly Labor Review*, December 1929, pp. 124–34, January 1930, pp. 118–26.

36. Bruce Smith, "What the Depression Has Done to Police Service," *Public Management*, March 1934, p. 67; Fogelson, *Big-City Police*, pp. 198, 210–12, 254–55, 278–79; International City Managers' Association, *The Municipal Year Book* (Chicago, 1940), pp. 394–95, 430–31, *The Municipal Year Book* (Chicago, 1960), pp. 364–65, 398–99; International City Management Association, *The Municipal Year Book 1979* (Washington, D.C., 1979), pp. 187–88; U.S. Bureau of Labor Statistics, *Salary Trends: Firemen and Police, 1924–64* (Washington, D.C., 1965). The earnings figures do not include overtime or longevity pay, which many cities add in when computing retirement allowances. If they were included, the rank and file's earnings would be a good deal higher. See Senate Subcommittee on Governmental Efficiency and the District of Columbia, *Hearings on the District of*

Columbia Pension Reform Act, pp. 44–53. Even higher would be the earnings of the superior officers, who make up anywhere from just over 10 to just under 40 percent of the big-city police forces. See John F. Heaphy, ed., *Police Practices: The General Administrative Survey* (Washington, D.C.: Police Foundation, 1978), p. 10.

37. Commonwealth of Massachusetts, *Report of the Commission on Pensions*, pp. 128–29, 233–46.

38. *Detroit News*, August 12, 1954; House Committee on the District of Columbia, *Hearings on Amendments to the Policemen's and Firemen's Retirement and Disability Act*, p. 5; U.S. Congress, House Committee on the District of Columbia, *Equalize Pensions of Retired Policemen and Firemen of the District of Columbia*, 67th Cong., 4th sess., H.R. 1400, pp. 1–2; *Police Pension and Relief Board of City and County of Denver* v. *McPhail*, 338 P.2d 694; Civic Federation of Chicago, *Bulletin No. 82*, April 1927, *Bulletin No. 95*, December 1928; Citizens Budget Commission, "Financial Aspects of New York City's Pension Systems" (1933), pp. 93–94; *Minneapolis Tribune*, February 23, 1969; Janet M. Corpus, "Dollar for Dollar: A Study of the Cost of Living Escalator in the Los Angeles Fire and Police Pension System," paper written for the Harvard–MIT Joint Center for Urban Studies' project on municipal employee unions (1977), chs. 2, 3; Harry Charles Katz, "The Impact of Public Employee Unions on City Budgeting and Employee Remuneration—A Case Study of San Francisco" (Ph.D. diss., University of California, Berkeley, 1977), pp. 130–31.

39. Letter to the author from Wm. S. Hutchison, Los Angeles Fire and Police Pension System, December 21, 1979; Senate Subcommittee on Governmental Efficiency and the District of Columbia, *Hearings on the District of Columbia Pension Reform Act*, pp. 39, 58–63.

40. Commission on Pensions, *Report on the Pension Funds of the City of New York*, pt. 1, table 22; Bureau of the Census, *Retirement Systems for State and Local Government Employees: 1941*, pp. 106–11; Bureau of the Census, *Employee Retirement Systems of State and Local Governments: 1972*, pp. 30–53.

41. Bureau of the Census, *Employee Retirement Systems of State and Local Governments: 1972*, pp. 30–53; Bureau of the Census, *Employee-Retirement Systems of State and Local Governments: 1977*, pp. 30–55; Senate Committee on the District of Columbia, *Hearings on Fiscal Pressures on the District of Columbia*, pp. 288–90; O'Brien letter, November 21, 1979; U.S. Bureau of the Census, *Finances of Employee-Retirement Systems of State and Local Governments in 1977–78* (Washington, D.C., 1979), p. 32; San Francisco City and County Employees' Retirement System, *Annual Report: 1976–1977*, p. 18. Figures are for 1977 in all cities except Washington, D.C., where they are for 1978.

42. Bureau of the Census, *Retirement Systems for State and Local Government Employees: 1941*, pp. 106–11; Bureau of the Census, *Employee Retirement Systems of State and Local Governments: 1972*, pp. 30–53; Bureau of the Census, *Employee-Retirement Systems of State and Local Governments: 1977*, pp. 30–55; Senate Committee on the District of Columbia, *Hearings on Fiscal Pressures on the District of Columbia*, pp. 291–92; O'Brien letter, November 21, 1979.

43. Commission on Pensions, *Report on the Pension Funds of the City of New*

York, pt. 1, pp. 165–66; *Report of the Illinois Pension Laws Commission*, pp. 11, 13; Bureau of the Census, *Financial Statistics of Cities: 1915*, pp. 190–91; Bureau of the Census, *Retirement Systems for State and Local Government Employees: 1941*, pp. 98–99, 106–11; Bureau of the Census, *Employee Retirement Systems of State and Local Governments: 1972*, pp. 30–53; Bureau of the Census, *Employee-Retirement Systems of State and Local Governments: 1977*, pp. 30–55.

44. Bureau of the Census, *Employee Retirement Systems of State and Local Governments: 1972*, pp. 30–53; Bureau of the Census, *Employee-Retirement Systems of State and Local Governments: 1977*, pp. 30–55.

45. Bureau of the Census, *Employee Retirement Systems of State and Local Governments: 1972*, pp. 30–53; Bureau of the Census, *Employee-Retirement Systems of State and Local Governments: 1977*, pp. 30–55; Senate Committee on the District of Columbia, *Hearings on Fiscal Pressures on the District of Columbia*, pp. 288–92; O'Brien letter, November 21, 1979; Los Angeles Board of Pension Commissioners, *Annual Report: 1978*, p. 42; Hutchison letter, December 21, 1979; Richard F. Camus and Associates, "Actuarial Study of the New Orleans Fire and Police Pension Funds," pp. 40F–43F, 20P–25P; Loretta Bevon and Louis D. Brown, "The Retirement Systems of the City of New Orleans," report prepared for the New Orleans Chief Administrative Officer, August 30, 1974, p. 7; Brune letter, March 11, 1980; Detroit Policemen and Firemen Retirement System, *Thirty-Seventh Annual Report: 1978*, p. 17; Soullier letter, December 20, 1979; telephone conversation with Tracy Howard, Denver Office of Budget and Management, December 2, 1980; San Francisco City and County Employees' Retirement System, *Annual Report: 1977–1978*, p. 21; Schwartz interview, November 26, 1979; Conrad M. Siegel, Inc., "Municipal Pension Funds," schedules 1PCI, 1FCI, 3PCI, 3FCI; telephone conversation with Walter L. Johnson, November 27, 1979; Coates, Herfurth & England, "Actuarial Valuation of the San Francisco Employees' Retirement System as of June 30, 1978," schedules 1, 3. Data are for 1977 except in the cases of New York, Detroit, Los Angeles, San Francisco, and Washington, D.C., where they are for 1978.

3. Lagging Revenues and Mounting Deficits

1. Chas. P. Neill, *Pension Funds for Municipal Employees and Railroad Pension Systems in the United States*, U.S. Congress, Senate Documents, 61st Cong., 2d sess., no. 427, pp. 36–85; Chicago Firemen's Pension Fund, *Report of Audit of Fund* (Chicago, 1919), pp. 6–11; U.S. Congress, House Committee on the District of Columbia, Subcommittee on Incorporations, *Report of Hearings on H.R. 22322* (Washington, D.C., 1910), p. 8; New York City Commission on Pensions, *Report on the Pension Funds of the City of New York* (New York, 1916), pt. 1, pp. 50–73, tables 54 and 55.

2. New York Bureau of Municipal Research, *A Report on the Police Pension Fund of the City of New York Submitted to the Aldermanic Committee on Police Investigation* (New York, 1913), pp. 16–19, 114, 116–17; Commission on Pensions,

Report on the Pension Funds of the City of New York, pt. 1, pp. 50–51, 63–67, tables 54 and 55.

3. House Committee on the District of Columbia, *Report of Hearings on H.R. 22322*, pp. 5–6, 35–36; J. D. Craig, *Report on the Firemen's Relief and Pension Fund, the Police Pension Fund and the Public School Teachers' Retirement Fund of the City of Buffalo* (1917), pp. 19, 38–39; *Report of the Illinois Pension Laws Commission* (Springfield, 1917), pp. 11–15, 272–73; *Report of the [New Jersey] Pension and Retirement Funds Commission* (1917), pp. 7–11; Commission on Pensions, *Report on the Pension Funds of the City of New York*, pt. 1, p. 50.

4. Paul Studensky, *The Pension Problem and the Philosophy of Contributions* (New York, 1917), pp. 1–2; Lawson Purdy, "Making Pension Systems an Asset," *American City*, January 1917, pp. 17–18; Municipal League of Los Angeles, *Bulletin*, November 12, 1917, pp. 5–6; Municipal League of Los Angeles to Los Angeles City Council, April 22, 1919, Los Angeles City Council Files, 1919, no. 867, Municipal Records Center, Los Angeles.

5. George B. Buck, "Municipal Pensions and Pension Funds," in National Association of Comptrollers and Accounting Officers, *Proceedings of the Tenth Annual Convention* (1915), pp. 105–10; George B. Buck, "An Actuarial Retirement System," *Good Government*, September 1918, pp. 137–41; Studensky, *The Pension Problem*, pp. 3–18; Paul Studensky, "Pensions in Public Employment," *National Municipal Review*, April 1922, pp. 103–8; Lewis Meriam, *Principles Governing the Retirement of Public Employees* (New York, 1918), pp. 92–98, 325–37.

6. Commonwealth of Massachusetts, *Report of the Commission on Pensions* (Boston, 1914), pp. 264–311; Commission on Pensions, *Report on the Pension Funds of the City of New York*, pt. 1, table 54; *Report of the Illinois Pension Laws Commission*, p. 172; U.S. Bureau of the Census, *Financial Statistics of Cities Having a Population of Over 30,000: 1915* (Washington, D.C., 1916), pp. 70, 92–93.

7. Commonwealth of Massachusetts, *Report of the Commission on Pensions*, p. 265; Buck, "Municipal Pensions and Pension Funds," p. 106; Commission on Pensions, *Report on the Pension Funds of the City of New York*, pt. 1, p. 76, pt. 2, pp. 62–63, 90–91; *Report of the Illinois Pension Laws Commission*, pp. 11–15.

8. Emily Jean Tourin, "The Politics of Pension Reform: New York State 1971–76" (Master's thesis, MIT, 1979), pp. 25–26.

9. Emma Schweppe, *The Firemen's and Patrolmen's Unions in the City of New York* (New York: Columbia University Press, 1948), chs. 2–6; Robert M. Fogelson, *Big-City Police* (Cambridge, Mass.: Harvard University Press, 1977), pp. 196–97; Philip Kenneth Kienast, "Policemen and Fire Fighter Employee Organizations" (Ph.D. diss., Michigan State University, 1972), chs. 2 and 4.

10. Schweppe, *Firemen's and Patrolmen's Unions*, pp. 186–87.

11. *Ibid.; International Fire Fighter*, November 1927, pp. 5–6.

12. Municipal League of Los Angeles, *Bulletin*, February 16, 1925, p. 12, May 1925, p. 10, August 8, 1925, pp. 5–6; *Los Angeles Times*, January 24, 1926; Los Angeles Fire and Police Protective League to Los Angeles City Council, March 25, 1925, Los Angeles City Council Files, 1925, no. 2090.

13. *Report of Pension Laws Commission of Milwaukee* (Milwaukee, 1920), pp.

23–62; Ovid B. Blix, "Pensions for Public Employees," *Milwaukee Government Service*, November 1, 1936, pp. 4, 12; *Milwaukee Sentinel*, August 7, 1918, November 28, 1920; *Wisconsin News*, January 27, 1921; *Milwaukee Journal*, June 5, 1921, August 12, 1923; *Milwaukee Leader*, June 2, 1921, December 26, 1923; Harold L. Henderson, "Milwaukee Adopts Scientific Pension System for Police Department," *American City*, September 1922, p. 263.

14. U.S. Bureau of the Census, *General Statistics of Cities: 1915* (Washington, D.C., 1916), pp. 96–97; U.S. Bureau of the Census, *Statistics of Fire Departments of Cities Having a Population of Over 30,000: 1917* (Washington, D.C., 1918), pp. 68–69; Citizens Budget Commission, *Compensation, Conditions of Employment and Retirement Benefits of Policemen and Firemen* (New York, 1938), pp. 24–25, 30–31.

15. *Detroit News*, January 23, 26, July 10, 11, 13, August 1, 2, 8, 14, September 16, October 24, 30, November 6, 14, 1940. For background, see *ibid.*, January 22, 1939; Detroit Bureau of Governmental Research, "The Pension Systems of the City of Detroit" (1929).

16. *Los Angeles Times*, May 22, 1947; Jack Allen to Los Angeles City Council, May 22, 1947, Los Angeles City Council Files, no. 27431; Los Angeles Board of Pension Commissioners, *Annual Report: 1978*, p. 25; Joseph Gerald Woods, "The Progressives and the Police: Urban Reform and the Professionalization of the Los Angeles Police" (Ph.D. diss., University of California, Los Angeles, 1973), pp. 392–96; Fogelson, *Big-City Police*, pp. 195–96.

17. U.S. Congress, House Committee on the District of Columbia, "Report to Accompany H.R. 6517," 85th Cong., 1st sess., H.R. 669, pp. 1–5; *Congressional Record*, May 22, 1957, pp. 7382–87, June 3, 1957, pp. 8197–210, August 10, 1957, pp. 14324–40. See also the unpublished transcripts of the hearings of the House Committee on the District of Columbia on H.R. 6517 and H.R. 7095 and the hearings of the Subcommittee on Fiscal Affairs of the Senate Committee on the District of Columbia on S. 1770, Senate District of Columbia Committee Files, National Archives, Washington, D.C.

18. *Detroit News*, October 28, 31, November 10, 1937; Schweppe, *Firemen's and Patrolmen's Unions*, pp. 191–92; *Milwaukee Journal*, July 17, 1938; Citizens Budget Commission, *Compensation, Conditions of Employment and Retirement Benefits*, pp. 24–25, 50–51; U.S. Congress, Senate Committee on Governmental Affairs, Subcommittee on Governmental Efficiency and the District of Columbia, *Hearings [on the] District of Columbia Pension Reform Act* (Washington, D.C. 1978), pp. 85–91.

19. U.S. Bureau of the Census, *Financing State and City Pensions* (Washington, D.C., 1941), pp. 51–59; U.S. Bureau of the Census, *Retirement Systems for State and Local Government Employees: 1941* (Washington, D.C., 1943), pp. 98–102; U.S. Bureau of the Census, *1957 Census of Governments: Volume IV, Topical Studies: Number 1, Employee-Retirement Systems of State and Local Governments* (Washington, D.C., 1959), pp. 30–49; U.S. Bureau of the Census, *1972 Census of Governments: Volume 6, Topical Studies: Number 1, Employee Retirement Systems of State and Local Governments* (Washington, D.C., 1973), pp. 30–53; U.S. Bureau of the Census, *Census of Governments, 1977: Volume 6, Topical Studies: Num-*

ber 1, Employee-Retirement Systems of State and Local Governments (Washington, D.C., 1978), pp. 30–55; letter to the author from Thomas J. O'Brien, Office of the City Administrator, Washington, D.C., November 21, 1979.

20. U.S. Department of Labor, Bureau of Labor Statistics, *Public Service Retirement Systems* (Washington, D.C., 1929), pp. 150, 169; Bureau of the Census, *Retirement Systems for State and Local Government Employees: 1941*, pp. 98–102; Bureau of the Census, *Employee-Retirement Systems of State and Local Governments: 1957*, pp. 30–49; Bureau of the Census, *Employee Retirement Systems of State and Local Governments: 1972*, pp. 30–53; Bureau of the Census, *Employee-Retirement Systems of State and Local Governments: 1977*, pp. 30–55.

21. Bureau of the Census, *Employee-Retirement Systems of State and Local Governments: 1977*, pp. 30–55; San Francisco City and County Employees' Retirement System, *Annual Report: 1978–1979*, pp. 45–47; O'Brien letter, November 21, 1979; interview with Jonathan Schwartz, New York City Employees' Retirement System, November 26, 1979. All figures are for 1977 except in the case of Washington, D.C., where they are for 1978, and New York, where, for the old systems, they are also for 1978.

22. "Analysis of Pensions" and "Fire and Police Pension Fund Revenues," Los Angeles City Council Files, no. 21072; Los Angeles Board of Pension Commissioners, Annual Reports, 1958–78; Detroit Policemen and Firemen Retirement System, *Thirty-Seventh Annual Report: 1978*, pp. 8–9; Oakland Police and Fire Retirement System, *Twenty-Sixth Annual Report: 1977*, pp. 25–27; U.S. Bureau of the Census, *Finances of Employee-Retirement Systems of State and Local Governments in 1977–78* (Washington, D.C., 1979), pp. 10, 30.

23. Washington, D.C., Office of Budget and Management Systems, "Retirement Facts: District of Columbia Police and Fire Retirement System" (1975), exhibit I; Ruth Ittner to Jack A. Cameron, January 31, 1968, Washington Public Pension Commission Files, Record Group 76, Division of Archives and Records Management, Department of General Administration, State of Washington, Olympia.

24. Economic Development Council of New York City, Inc., "Pension Changes in New York City, 1962–72" (1972), pp. 44, 47; Harry Charles Katz, "The Impact of Public Employee Unions on City Budgeting and Employee Remuneration—A Case Study of San Francisco" (Ph.D. diss., University of California, Berkeley, 1977), pp. 128–29; U.S. Congress, Senate Committee on the District of Columbia, *Hearings [on] Fiscal Pressures on the District of Columbia: Part 3. Pension Systems* (Washington, D.C., 1976), pp. 439–42.

25. *Minneapolis Tribune*, February 23, 1969; Minnesota State Planning Agency, Office of Local and Urban Affairs, *Minneapolis–St. Paul Study. Part II: Report on Pensions* (1978), p. 38; International City Managers' Association, *The Municipal Year Book* (Chicago, 1967), pp. 397, 454; Janet M. Corpus, "Dollar for Dollar: A Study of the Cost of Living Escalator in the Los Angeles Fire and Police Pension System," paper written for the Harvard–MIT Joint Center for Urban Studies' project on municipal employee unions (1977), ch. 3; Los Angeles Board of Pension Commissioners, *Annual Report: 1978*, p. 31; Katz, "Impact of Public Employee Unions," pp. 130–31; *San Francisco Chronicle*, October 16, 1974; *San*

Francisco Examiner, October 28, 1974; San Francisco City and County Employees' Retirement System, *Annual Report: 1974–1975*, p. 13, *1975–1976*, p. 14.

26. Senate Committee on Governmental Affairs, *Hearings on the District of Columbia Pension Reform Act*, pp. 44–53; Kansas City Police Department, "General Administrative Survey" (1970), table 2. The Los Angeles salary data were compiled from *The Municipal Year Book*. The figures assume that the officer reached maximum grade after five years of service. They do not include overtime, longevity pay, or other increments. The results would have been the same for a Los Angeles fire fighter.

27. *Milwaukee Journal*, July 17, 1978; Temporary Commission on City Finances, *The City in Transition: Prospects and Policies for New York* (New York, 1977), pp. 181–82; "Pensions: A Report of the [New York City] Mayor's Management Advisory Board" (1976), pp. 53–54.

28. *Report of the Illinois Pension Laws Commission*, pt. 2, chs. 2 and 3; Retirement Board of Policemen's Annuity and Benefit Fund of Chicago, *Report for the Calendar Year 1922*, pp. 2–11; *Chicago Tribune*, April 3, 5, June 20, 1921.

29. Anthony Pratt to Los Angeles City Council, August 1, 1921, Los Angeles City Council Files, 1921, no. 2146; Municipal League of Los Angeles, *Bulletin*, February 16, 1925, p. 12; *Los Angeles Herald*, June 16, 1922; *Los Angeles Times*, March 25, 1922, August 31, December 6, 1931, February 11, 15, 21, March 25, April 26, 1932; Los Angeles Fire and Police Protective League, *History of the Department of Pensions of the City of Los Angeles* (1939), pp. 4–8.

30. Citizens Budget Commission, "Financial Aspects of New York City's Pension Systems" (1933), pp. 88–100; Harold Riegelman, *Number One Municipal Problem: The Unsound Pension Funds* (New York, 1940), pp. 4–6; New York State Commission on Pensions, "Report of an Actuarial Investigation of the N.Y. Fire Department Relief Fund as of December 31, 1933" (1934), p. 15; New York State Commission on Pensions, "Report of an Actuarial Investigation of the Police Pension Fund of the City of New York as of December 31, 1933" (1934), p. 15; Schweppe, *Firemen's and Patrolmen's Unions*, pp. 185–88.

31. "New Pension Plan for Municipal Employees Effective in Cincinnati," *American City*, November 1931, pp. 95–96; Ralph R. Nelson, "San Francisco City and County Employees' Retirement Systems," memorandum dated October 28, 1957, San Francisco City and County Employees' Retirement System Files, San Francisco; *Detroit News*, March 5, October 3, 1937; *Greater Cleveland*, April 11, 1928, pp. 129–40; International City Managers' Association, *The Municipal Year Book 1938* (Chicago, 1938), pp. 318–21.

32. *Los Angeles Examiner*, January 28, 1954; M. M. Devore to Norris Poulson, January 24, 1957, James Cambray to Norris Poulson, January 16, 1957, Los Angeles City Council Files, no. 77868; Samuel Leask, Jr., to Charter and Administrative Code Committee, December 1, 1958, Los Angeles City Council Files, no. 77868, supp. 3; Ilene G. Greenberg, "Going Through the Motions: A Study of Pension Reform in Los Angeles," paper written for the Harvard–MIT Joint Center for Urban Studies' project on municipal employee unions (1977), pp. 7–20.

33. Minnesota Public Retirement Systems Interim Commission, *Report to the*

1967 Legislative Session of the State of Minnesota (1967), pp. 29–44; R. E. Curran, "Our Position on Pensions," *Minnesota Fire Chief,* January–February 1967, pp. 25–26; "Important Changes Proposed for Police and Firemen's Pensions," *Minnesota Municipalities,* February 1967, pp. 40–42; Orville C. Peterson, "Proposed Statute on Police-Firemen's Pension Plans," memorandum dated December 4, 1968, Minnesota Legislative Commission on Pensions and Retirement Files, St. Paul; *Minneapolis Tribune,* February 23, 1969; Minnesota Legislative Commission on Pensions and Retirement, *Report to the 1977–1978 Minnesota State Legislature,* pp. 67–69, 75.

34. Senate Committee on the District of Columbia, *Hearings on Fiscal Pressures on the District of Columbia,* pp. 285–311; Senate Subcommittee on Governmental Efficiency and the District of Columbia, *Hearings on the District of Columbia Pension Reform Act,* pp. 4–16, 122–27, 151–54, 179–85, 691; *Washington Post,* April 7, 26, September 27, 1977, May 3, August 25, September 22, October 14, 1978, October 30, 1979; *Congressional Quarterly,* November 11, 1978, p. 3282.

35. Schweppe, *Firemen's and Patrolmen's Union,* pp. 187–93; Ohio Legislative Service Commission, *Funding of the Police and Firemen's Disability and Pension Fund* (Columbus, Ohio, 1969), p. 17; Senate Subcommittee on Governmental Efficiency and the District of Columbia, *Hearings on the District of Columbia Pension Reform Act,* pp. 85–91.

36. Bureau of the Census, *Retirement Systems for State and Local Government Employees: 1941,* pp. 98–99; Bureau of the Census, *Finances of Employee-Retirement Systems of State and Local Governments in 1977–78,* pp. 10–29; U.S. Treasury Department, "Actuarial Study of the Policemen's and Firemen's Relief Fund of the District of Columbia" (1942), a copy of which was made available to me by Ray Schmidt of the Congressional Research Service of the Library of Congress; O'Brien letter, November 21, 1979; *Portland Oregonian,* September 11, 1976; *Denver Sun,* February 12, 1978; Martin E. Segal Company, "Colorado Police & Firemen's Pension Plans Actuarial Valuations as of January 1, 1978" (1979), pp. 39, 44, 60, 65; telephone conversation with Robert P. Clohessy, Portland Fire and Police Disability and Retirement System, February 25, 1980. The projection for Denver was for the police pension fund only. Government contributions include state as well as city contributions.

37. Los Angeles Board of Pension Commissioners, *Annual Report: 1959,* pp. 19, 25, *1978,* pp. 24, 42; Detroit Auditor General, *Annual Report: 1943,* p. 26, *1978,* exhibits R-1, R-2; Bureau of the Census, *Finances of Employee-Retirement Systems of State and Local Governments in 1977–78,* pp. 10–29.

38. Detroit Auditor General, *Annual Report: 1943,* p. 26, *1978,* exhibits R-1, R-2; Los Angeles Board of Pension Commissioners, *Annual Report: 1959,* pp. 19, 25, *1978,* pp. 24, 42; Bureau of the Census, *Finances of Employee-Retirement Systems of State and Local Governments in 1977–78,* pp. 10–29.

39. *Report of the Illinois Pension Laws Commission,* pp. 11–15, 101, 121; telephone conversation with William McGlone, Chicago Civic Federation, February 25, 1980; Conrad M. Siegel, Inc., "Act 293 Report. Municipal Pension Funds. 1976 Actuarial Study Analysis," report prepared for the Pennsylvania Department of Community Affairs (1977), schedules 1PCI, 1FCI, 3PCI, 3FCI.

40. Los Angeles Board of Pension Commissioners, *Annual Report: 1960*, pp. 6, 29, *1978*, pp. 6, 7, 30; Los Angeles Department of Pensions, "Fire and Police Pension System" (1975), chart facing p. 9, Department of Pensions Files, Los Angeles; letter to the author from Wm. S. Hutchison, Los Angeles Fire and Police Pension System, February 21, 1979; Los Angeles Fire and Police Protective League, *History of the Department of Pensions*, p. 8; Detroit Auditor General, *Annual Report: 1942*, p. 29; Bureau of the Census, *Retirement Systems for State and Local Government Employees: 1941*, p. 106; Detroit Policemen and Firemen Retirement System, *Thirty-Seventh Annual Report: 1978*, pp. 5–6; Schwartz interview, November 26, 1979; Coates, Herfurth, & England, "Actuarial Valuation of the San Francisco City and County Employees' Retirement System as of June 30, 1978" (1979), p. 6, schedules 1-a, 1-b, 1-c; telephone conversation with Walter L. Johnson, Oakland Municipal Employees' Retirement System, December 18, 1980. Oakland's old fire and police pension system, a pay-as-you-go plan that was closed to new members in 1951, had another $200 million or so in unfunded liabilities. See Oakland Police and Fire Retirement System, *Twenty-Sixth Annual Report: 1977*, pp. 12–13. Minnesota Legislative Commission on Pensions and Retirement, *Report to the 1979–1980 Minnesota State Legislature*, p. 47; Conrad M. Siegel, Inc., "Municipal Pension Funds," schedules 1NCI, 3NCI; "Report of the Mayor's Management Advisory Board," pp. 16, 19, 50; New Orleans Employees' Retirement System, *Twenty-Ninth Fiscal Report* (1975), pp. 8, 10; Gabriel, Roeder, Smith & Company, "The Report of the 37th Annual Actuarial Valuations June 30, 1975 of the City of Detroit General Retirement System" (1976), pp. 6, 10; Comptroller General of the United States, "Funding of State and Local Government Pension Plans: A National Problem" (1979), pp. 38, 42, 51, 55.

41. Los Angeles Board of Pension Commissioners, *Annual Report: 1960*, p. 4; H. P. Dowling to Detroit City Council, June 19, 1959, Detroit City Council Files, Detroit; telephone conversation with Edmond J. Walsh, San Francisco City and County Employees' Retirement System, March 7, 1980; District of Columbia Retirement Reform Act, Public Law 96-112, November 17, 1979, sections 142, 144, 145.

42. Milliman & Robertson, "Actuarial Report and Recommended Contribution Rates to the Retirement Board Oakland Police and Fire Retirement System as of 30 June 1975" (1976), pp. 1–14. For a defense of the actuarial practices criticized by Milliman & Robertson, see Coates, Herfurth & England, "A Report on the Oakland Police and Fire Retirement System" (1976), pp. 3–16.

43. *Detroit News*, December 13, 1976; Minnesota State Planning Agency, *Report on Pensions*, pp. 42–43; Robert Tilove, *Public Employee Pension Funds* (New York: Columbia University Press, 1976), pp. 281–82.

44. *Seattle Post-Intelligencer*, January 24, 1977.

45. Thomas P. Bleakney, "Washington Law Enforcement Officers' and Firefighters' Retirement System Second Actuarial Valuation" (1971), p. 5; Milliman & Robertson, "Washington Law Enforcement Officers' and Firefighters' Retirement System Actuarial Valuation as of December 31, 1975" (1976), pp. 9, 46–47; *Seattle Post-Intelligencer*, January 26, 1977.

46. *Tomlinson* v. *Kansas City*, 391 S.W.2d. 850; Judy A. Levenson, "The Al-

batross: An Essay on the Detroit Policemen's and Firemen's Retirement System, 1968," paper written for the Harvard–MIT Joint Center for Urban Studies' project on municipal employee unions (1977), pp. 15–16; Los Angeles Board of Pension Commissioners, *Annual Report: 1967*, pp. 4–5; Tilove, *Public Employee Pension Funds*, pp. 282–83.

47. Steve Brandt, "Public Pensions: The Sorry State of Our State," *Boston Globe*, March 9, 1980.

4. The Politics of Pension Reform

1. *Portland City Club Bulletin*, December 2, 1932, p. 3, December 9, 1932, pp. 1, 3, 4; Herman Kehrli, *Minneapolis Pension Systems* (Minneapolis, 1933), foreword and introductory.

2. Kehrli, *Minneapolis Pension Systems*, foreword; Harold Riegelman, *Number One Municipal Problem: The Unsound Pension Funds* (New York, 1939), p. 11; Anthony Pratt to Los Angeles City Council, June 23, 1921, Los Angeles City Council Files, 1922, no. 2404, Municipal Records Center, Los Angeles.

3. Herman Kehrli, "Portland Pension Problems," report to the Portland City Council (1934), p. 3; New Orleans Bureau of Governmental Research, *City Problems*, May 22, 1944, pp. 423–30; *Greater Cleveland*, October 21, 1943, pp. 21–28; Citizens Budget Commission, "Financial Aspects of New York City's Pension Systems" (1933), pp. 94–96, 99.

4. Municipal League of Los Angeles, *Bulletin*, August 28, 1925, pp. 5–6; Citizens Budget Commission, "New York City's Pension Systems," pp. 95–96, 99–100; *Greater Cleveland*, April 11, 1928, pp. 129–40, October 21, 1943, pp. 21–28; New Orleans Bureau of Governmental Research, *City Problems*, May 22, 1944, pp. 423–30; Kehrli, "Portland Pension Problems," pp. 92–94; Philadelphia Bureau of Municipal Research, "Report on the Philadelphia Police Pension Fund" (1953), pp. 29–31, "Report on the Philadelphia Firemen's Pension Fund" (1953), pp. 25–28.

5. Paul Studensky, "Pensions in Public Employment," *National Municipal Review*, April 1922, pp. 121–22; Paul Studensky, "Boston's Worthy Pension Endeavor," *National Municipal Review*, August 1921, pp. 406–7; "New Pension Plan for Municipal Employees Effective in Cincinnati," *American City*, November 1931, pp. 95–96; *Milwaukee Journal*, March 21, May 7, June 20, July 25, 1947; Citizens' Bureau of Milwaukee, "How Other Cities Established a Single Pension System for Firemen, Policemen, and General City Employees" (1945).

6. Municipal League of Los Angeles, *Bulletin*, August 28, 1925, pp. 5–6; Tom Ingersoll to Los Angeles Board of Pension Commissioners, November 30, 1931, Department of Pensions Files, Los Angeles; *Los Angeles Times*, December 6, 1931, March 25, 1932; Los Angeles Chamber of Commerce, State and Local Government Committee, "Proposed City of Los Angeles Charter Amendment No. 1" (1959), pp. 4–5; Janet M. Corpus, "Dollar for Dollar: A Study of the Cost of Living Escalator in the Los Angeles Fire and Police Pension System," paper written for the Harvard–MIT Joint Center for Urban Studies' project on municipal em-

ployee unions (1977), pp. 10–11; Ilene G. Greenberg, "Going Through the Motions: A Study of Pension Reform in Los Angeles," paper written for the Harvard–MIT Joint Center for Urban Studies' project on municipal employee unions (1977), pp. 12–16.

7. Emma Schweppe, *The Firemen's and Patrolmen's Unions in the City of New York* (New York: Columbia University Press, 1948), pp. 184–90; Riegelman, *Number One Municipal Problem*, pp. 4–7, 11–12; *Detroit News*, January 20, 1930, October 11, 1933, July 5, December 7, 1934, July 11, 1935, February 6, 1936, March 5, 1937; New Orleans Bureau of Governmental Research, *City Problems*, May 22, 1944, pp. 423–30; New Orleans Bureau of Governmental Research, *A Summary of the Bureau's Recommendations for Improved Government in New Orleans* (New Orleans, 1946), ch. 3, pp. 93–100.

8. Los Angeles Chamber of Commerce, State and Local Government Committee, "Proposed City of Los Angeles Charter Amendment No. 1," pp. 1–5; Los Angeles City Administrative Officer, "Review of Fire and Police Pension System," minutes of meetings of December 17 and December 27, 1963, City Administrative Officer's Files, Los Angeles; Walter C. Peterson to Los Angeles City Attorney, March 17, 1959; Los Angeles Fire and Police Protective League, "Statement of Policy," March 9, 1959; Samuel Leask, Jr., to Personnel Committee of City Council, March 9, 1959; Los Angeles Fire and Police Protective League, "Argument in Favor of Proposed Charter Amendment No. 1," Los Angeles City Council Files, no. 116278.

9. Los Angeles City Administrative Officer, "Review of Fire and Police Pension System," minutes of meetings of December 3, 4, 5, 10, 15, and 27, 1963, City Administrative Officer's Files; C. Erwin Piper to Los Angeles City Council, June 8, 1964, app., pp. 1–22, Los Angeles City Council Files, no. 116278.

10. C. Erwin Piper to Los Angeles City Council, "Proposed New Fire and Police Pension System," November 24, 1965, pp. 3–10, Los Angeles City Council Files, no. 116278. Under no circumstances would members have had to contribute more than 2 percent of salary for the cost-of-living adjustment.

11. *Los Angeles Herald-Examiner*, December 6, 1975; John A. Thompson to Los Angeles City Council, June 14, 1966, Los Angeles City Council Files, no. 130081; "Pension Report," *Los Angeles Fire and Police Protective League News*, March 1966, pp. 1–6; Greenberg, "Going Through the Motions," pp. 46–48.

12. C. Erwin Piper, "Proposed New Fire and Police Pension System," July 13, 1966, pp. 3–10; Los Angeles Chamber of Commerce, State and Local Government Committee, "Proposed New Fire and Police Pension System," August 4, 1966, pp. 1–3, Los Angeles City Council Files, no. 116278; *Los Angeles Herald-Examiner*, August 9, 1966; *Van Nuys News*, August 11, 1966; Greenberg, "Going Through the Motions," pp. 48–52.

13. *Los Angeles Times*, March 13, 1972; Greenberg, "Going Through the Motions," pp. 11–12, 52–55; Corpus, "Dollar for Dollar," pp. 24–25.

14. *Los Angeles Times*, August 23, 1966; "Proposition P," *Los Angeles Fire and Police Protective League News*, October 1966, p. 1; Melvin Horton, secretary of Property Owners Tax Association of California, transcript of broadcast on KNXT, October 13 and 14, 1966; "Vote Yes on P," a Los Angeles Fire and Police Pro-

tective League broadside, City Administrative Officer's Files; "Argument in Favor of Proposed Charter Amendment 'P,' " Election Division Files, City Clerk's Office, Los Angeles; Greenberg, "Going Through the Motions," pp. 58–65.

15. C. Erwin Piper to Sam Yorty, January 29, 1969, City Administrative Officer's Files; "Argument in Favor of Proposed Charter Amendment Number 2," Election Division Files; Otho R. Allen to Los Angeles City Council, March 9, 1970, Los Angeles City Council Files, no. 70-5576; Milton Harker, secretary of the Property Owners Tax Association of California, transcript of broadcast on KGIB, May 2, 1971; Los Angeles Chamber of Commerce, State and Local Government Committee, "Fire and Police Pension Charter Amendment Proposition 2, May 25, 1971," May 6, 1971, City Administrative Officer's Files; Los Angeles Fire and Police Protective League, "Official Audit and Report for the Campaign to Pass Charter Amendment 2" (1971).

16. Riegelman, *Number One Municipal Problem*, pp. 10–11.

17. *New York Times*, July 6, August 12, November 17, 1939; Schweppe, *Firemen's and Patrolmen's Unions*, pp. 187–92; John Tessler, "Pension Forum, Inc. A Case Study of an Insurgent Group in the New York City Police Department" (Master's thesis, CCNY, 1959), pp. 11–13.

18. *New York Times*, November 18, 28, December 28, 1939; Riegelman, *Number One Municipal Problem*, pp. 12–16; Schweppe, *Firemen's and Patrolmen's Unions*, pp. 192–93; Tessler, "Pension Forum, Inc.," pp. 13–14, 18–21.

19. Tessler, "Pension Forum, Inc.," pp. 22–25.

20. *Ibid.*, pp. 25–53.

21. *Ibid.*, pp. 37–106; *New York Times*, March 6, 1950.

22. Tessler, "Pension Forum, Inc.," pp. 102–10.

23. *New York Times*, October 3, 1950; *Civil Service Leader*, July 12, 1950, p. 14; "Statement of Daniel L. Kurshan, Executive Director of the Citizens Budget Commission on Police and Fire Pension Bills," statement to the Finance Committee of the City Council, October 10, 1950, Citizens Budget Commission Files, New York City.

24. Tessler, "Pension Forum, Inc.," pp. 110–30; Robert M. Fogelson, *Big-City Police* (Cambridge, Mass.; Harvard University Press, 1967), p. 204; *Civil Service Leader*, October 24, 1950, p. 1.

25. *New York Times*, August 8, 16, 17, 30, 1951; *Civil Service Leader*, April 8, 1950, p. 15; Tessler, "Pension Forum, Inc.," pp. 130–33; Economic Development Council of New York City, Inc., "Pension Changes in New York City, 1962–72" (1972), pp. 44–48.

26. A. A. Weinberg, "The Public Employee Pension Problem in the State of Washington," report of the Washington Legislative Council (1962), pp. 117–21; *Seattle Times*, May 3, 1962. For background, see Municipal League of Seattle and King County, *Municipal News*, January 28, 1955, pp. 17–18, January 24, 1959, pp. 9–10, February 11, 1961, pp. 17–18, February 25, 1961, pp. 25, 26, 32.

27. Weinberg, "Public Employee Pension Problem," pp. 215–24, 233–35.

28. A. A. Weinberg to Donald C. Sampson, April 29, 1962, October 15, 1962, A. A. Weinberg to Chester Biesen, October 29, 1962, Donald C. Sampson to

A. A. Weinberg, October 22, 1962, Washington Public Pension Commission Files, Record Group 76, Division of Archives and Records Management, Department of General Administration, State of Washington, Olympia.

29. *Seattle Times*, January 15, 1967, November 24, 1968.

30. C. A. Crosser to Homer Humiston, January 18, 1967, Thomas P. Bleakney to John S. Murray, February 2, 1967; F. C. Ramon to Public Pension Commission, August 16, 1966, Gordon F. Vickery to Public Pension Commission, August 19, 1966, Homer Humiston, "Memorandum on House Bill 113," January 30, 1967, Public Pension Commission Files; *Tacoma News Tribune*, January 20, 1967.

31. Homer Humiston to Seattle Fire Department Personnel, memorandum dated March 25, 1967; Public Pension Commission, "Comparison of Benefits for Firemen and First-Class City Policemen with Benefits Proposed by H.B. 950 (1967) for All Policemen and Firemen," table prepared in November 1967, Public Pension Commission Files.

32. Floyd C. Miller to Public Pension Commission, May 21, 1968, Thomas P. Bleakney to Chester Biesen, December 26, 1967, Gordon L. Vickery to Public Pension Commission, January 31, 1968, Public Pension Commission Files.

33. Homer Humiston to Seattle Fire Department Personnel, memorandum dated March 25, 1967, Homer Humiston to John Smith, Jr. March 25, 1967, Public Pension Commission Files; *[Seattle] Fire Fighter*, March 1968, pp. 5–6.

34. "Statement by Homer Humiston on House Bill 353 Before House Committee on Labor, February 27, 1969," pp. 1–4, Public Pension Commission Files; *The Guardian*, a publication of the Seattle Police Officers' Guild, May 1977, p. 3; *Spokane Chronicle*, July 9, 1969; *Seattle Post-Intelligencer*, January 9, 1977.

35. *Seattle Post-Intelligencer*, January 9, 23–27, 1977; *Seattle Times*, April 7, 14, 1974, January 29, 1977, February 13, 1978.

36. Judy A. Levenson, "The Albatross: An Essay on the Detroit Policemen's and Firemen's Retirement System, 1968," paper written for the Harvard–MIT Joint Center for Urban Studies' project on municipal employee unions (1977), pp. 8–15; Public Service Administration, "Compensation and Related Practices in Major United States Cities for Police and Fire Services" (1963), pp. 51–72.

37. Levenson, "The Albatross," pp. 15–18; *Detroit Free Press*, July 30, 1966; Detroit Auditor General, *Annual Report: 1967*, p. 15, *1968*, p. 15; Detroit Policemen and Firemen Retirement System, *Twenty-Seventh Annual Report: 1968*, schedule 13.

38. Levenson, "The Albatross," pp. 18–20; Margaret A. Levi, *Bureaucratic Insurgency* (Lexington, Mass.: Heath, 1977), pp. 100–2.

39. Levenson, "The Albatross," pp. 20–25.

40. *Ibid.*, pp. 25–30; *Detroit News*, September 2–4, 1968; *Detroit Free Press*, August 26, September 4, 1968.

41. *Detroit Free Press*, August 26, 1968.

42. *Ibid.*, October 1, 6, 1968; Levenson, "The Albatross," p. 31.

43. Levenson, "The Albatross," pp. 30–37; *Detroit Free Press*, October 6, 10, 27, 29, 1968; *Detroit News*, September 9, October 1, 25, 27, 30, November 3, 4, 1968.

44. Levenson, "The Albatross," pp. 38–52.

45. *Los Angeles Times*, March 13, 1972; *New Orleans States-Item*, May 3, 1967; Ken Auletta, *The Streets Were Paved With Gold* (New York: Random House, 1979), p. 299.

46. Fogelson, *Big-City Police*, ch. 8; Sterling D. Spero and John M. Capozzola, *The Urban Community and Its Unionized Bureaucracies* (New York: Dunellen, 1973), ch. 3.

47. *Seattle Post-Intelligencer*, January 9, 1977; Auletta, *The Streets Were Paved with Gold*, p. 325; Nancy Lombardo, "Money, Money, Money: Public Unions and Campaign Finance in Boston," paper written for the Harvard–MIT Joint Center for Urban Studies' project on municipal employee unions (1977).

48. *Seattle Times*, April 7, 1974.

49. David Lewin, "Wage Determination in Local Government Employment" (Ph.D. diss., University of California, Los Angeles, 1971), pp. 123–25; Donald Whitney Berney, "Law and Order Politics: A History and Role Analysis of Police Officer Organizations" (Ph.D. diss., University of Washington, 1971), ch. 5.

50. Fogelson, *Big-City Police*, chs. 2 and 6.

51. Horton, "Vote Yes on P"; *Detroit Free Press*, October 16, 1968.

52. Richard Rovere, "Letter from Washington," *The New Yorker*, November 14, 1977, p. 203.

5. *The Legal Constraints on Reform*

1. Harry Charles Katz, "The Impact of Public Employee Unions on City Budgeting and Employee Remuneration—A Case Study of San Francisco" (Ph.D. diss., University of California, Berkeley, 1977), pp. 131–32; Office of the State Actuary, "Law Enforcement Officers and Firefighters Retirement System of the State of Washington Actuarial Valuation 1978" (1980), pp. 44–51; Dick Brown, "Policemen's and Firemen's Pensions: 1978 and 1979 Legislative Reforms" (1979), pp. 7–10.

2. *Eddy* v. *Morgan*, 216 Ill. 437 (1905), quote on p. 449; *Blough* v. *Ekstrom*, 144 N.E. 2d 436 (1957); *Pecoy* v. *City of Chicago*, 106 N.E. 435 (1914); *MacFarland* v. *Bieber*, 32 App. D.C. 513 (1909), quotes on pp. 520–21; *Pennie* v. *Reis*, 132 U.S. 464 (1889), quotes on p. 471; *Gibbs* v. *Minneapolis Fire Dep't. Relief Ass'n.*, 145 N.W. 1075 (1914), quote on p. 1076; *City of Dallas* v. *Trammell*, 101 S.W.2d 1009 (1937), quote on p. 1017. For a compilation and discussion of cases on the subject, see Annot., Vested Right of Pensioner to Pension, 54 A.L.R. 943 (1928).

3. *Rudolph* v. *United States*, 36 App. D.C. 379 (1911), quotes on pp. 384–85; *Pennie* v. *Reis*, p. 471; *Pecoy* v. *City of Chicago*, p. 436; *Eddy* v. *Morgan*, p. 449; *Gibbs* v. *Minneapolis Fire Dep't. Relief Ass'n; People ex rel. Donovan* v. *Retirement Board Policemen's Annuity and Benefit Fund*, 158 N.E. 220 (1927).

4. *Beutel* v. *Foreman*, 288 Ill. 107 (1919); *Lickert* v. *City of Omaha*, 12 N.W.2d 644 (1944); *MacFarland* v. *Bieber; Board of Trustees of Firemen's Pension Fund for*

City & County of Denver v. *People ex rel. Behrman*, 203 P.2d 480 (1949); *State ex rel. Holton* v. *City of Tampa*, 119 Fla. 556; *City of Dallas* v. *Trammell*.

5. *O'Dea* v. *Cook*, 169 P. 366 (1917), quote on p. 367; *Retirement Board of Allegheny County* v. *McGovern*, 174 A. 400 (1934), quote on p. 404; *Adamson* v. *City of Little Rock*, 134 S.W.2d 558 (1939); *Bakenhus* v. *City of Seattle*, 296 P.2d 536 (1956); *Hickey* v. *Pension Board of City of Pittburgh*, 106 A.2d 233 (1954), quote on p. 235; *Dryden* v. *Board of Pension Commissioners of the City of Los Angeles*, 6 Cal. 2d 575 (1936); *Trotzier* v. *McElroy*, 186 S.E. 817 (1936); *Yeazell* v. *Copins*, 402 P.2d 541 (1965). For a compilation and discussion of cases on the subject, see Annot., Vested Right of Pensioner to Pension, 52 A.L.R.2d 437 (1957).

6. *Retirement Board of Allegheny County* v. *McGovern*, quotes on pp. 404–5; *Roddy* v. *Valentine*, 197 N.E. 260 (1935), quote on p. 262.

7. *Bowen* v. *City of Los Angeles*, 257 P.2d 672 (1953); *Hickey* v. *Pension Board of City of Pittsburgh*, quote on p. 237; *Bakenhus* v. *City of Seattle*, quote on p. 540.

8. *Trotzier* v. *McElroy*; *Bender* v. *Anglin*, 60 S.E.2d 756 (1950); *Police Pension and Relief Board of City and County of Denver* v. *McPhail*, 338 P.2d 694 (1959); *Police Pension and Relief Board of City and County of Denver* v. *Bills*, 366 P.2d 581 (1961); *Allen* v. *City of Long Beach*, 287 P.2d 765 (1955); *Bakenhus* v. *City of Seattle*; *Yeazell* v. *Copins*.

9. *Creps* v. *Firemen's Relief & Retirement Fund Trustees of Amarillo*, 456 S.W.2d 434 (1970); *Spina* v. *Consolidated Police and Firemen's Pension Fund Commission*, 197 A.2d 169 (1964); *Dougherty* v. *United States ex rel. Browning*, 45 F.2d 926 (1930); U.S. Congress, House Committee on Education and Labor, *Pension Task Force Report on Public Employee Retirement Systems* (Washington, D.C., 1978), pp. 7–13.

10. Emma Schweppe, *The Firemen's and Patrolmen's Unions in the City of New York* (New York: Columbia University Press, 1948), pp. 185–87.

11. *Roddy* v. *Valentine*. See also House Committee on Education and Labor, *Pension Task Force Report*, p. 596.

12. Vernon A. O'Rourke and Douglas W. Campbell, *Constitution-Making in a Democracy* (Baltimore: Johns Hopkins University Press, 1943), pp. 59–60, 62–65: Schweppe, *Firemen's and Patrolmen's Unions*, p. 188; *Revised Record of the Constitutional Convention of the State of New York April Fifth to August Twenty-Sixth 1938*, vol. 2 (Albany, 1938), pp. 1404–6.

13. Harold Riegelman, *Number One Municipal Problem: The Unsound Pension Funds* (New York, 1939), p. 9; *Revised Record of the Constitutional Convention*, vol. 2, pp. 1406–11, 1414–18, vol. 3, pp. 2549–52, vol. 4, pp. 2984–86.

14. *Revised Record of the Constitutional Convention*, vol. 2, pp. 1412–14, 1419–21, vol. 3, pp. 2546–47.

15. *Ibid.*, vol. 2, pp. 1406–23.

16. *Ibid.*, vol. 3, pp. 2547–50; O'Rourke and Day, *Constitution-Making in a Democracy*, pp. 159–60.

17. *New York Times*, August 15, 16, 1938; *Revised Record of the Constitutional Convention*, vol. 4, pp. 2983–93.

18. O'Rourke and Day, *Constitution-Making in a Democracy*, ch. 8. See Henry J. Amy's letter to the editor, *New York Sun*, October 10, 1938, and the transcript of his address on WNYC, October 25, 1938, Citizens Budget Commission Files, New York City.

19. Austin Knapp, ed., *State of Michigan, Constitutional Convention, 1961, Official Record*, vol. 1, pp. 770–75; *Record of Proceedings, Sixth Illinois Constitutional Convention: Verbatim Transcripts* (1970), pp. 2929–33; Loren Oury, "Public Employee Pension Rights and the 1970 Illinois Constitution," *John Marshall Journal of Practice and Procedure*, Winter 1975–76, pp. 445–52; Louisiana Constitutional Convention Records Commission, *Records of the Louisiana Convention of 1973: Convention Transcripts* (1977), vol. 9, pp. 2564–86; House Committee on Education and Labor, *Pension Task Force Report*, pp. 401–10; *State of New Jersey Constitutional Convention of 1947*, vol. 3, pp. 103–7, 192, 427–28, vol. 5, pp. 494–97.

20. *Milwaukee Journal*, January 21, 1945; *Milwaukee Sentinel*, February 2, 3, 4, 8, March 29, 1945; *State ex rel. Bartelt* v. *Thompson*, 16 N.W.2d 420 (1944); *State ex rel. Risch* v. *Board of Trustees of Policemen's Pension Fund*, 98 N.W. 954 (1904); *State ex rel. McCarty* v. *Gantter*, 4 N.W.2d 153 (1942). The Wisconsin Supreme Court treated teachers differently, holding that their pensions were contractual obligations. For the court's reasoning, see *State ex rel. O'Neil* v. *Blied*, 206 N.W. 213 (1925).

21. *Milwaukee Sentinel*, January 31, February 2, 4, 9, 12, 14, 16, 25, 1945.

22. *Ibid.*, March 20–25, 1945; *Milwaukee Journal*, March 20, 21, 1945.

23. *Milwaukee Sentinel*, March 22, April 1–3, 5, 10, 17, 24, May 1, 3, 11, 1945; *Milwaukee Journal*, March 26, 29, April 6, 1945.

24. *Milwaukee Sentinel*, May 20, June 1, 13, 15, October 11, 16, 1945, December 11, 1946; *Milwaukee Journal*, May 20, 24, October 26, 1945, March 29, 1947.

25. *Milwaukee Journal*, October 25, 1945, December 11, 1946, March 17, 20, May 7, 1947; *Milwaukee Sentinel*, December 11, 1946, March 17, 21, 1947.

26. Milwaukee Citizens' Governmental Research Bureau, *Bulletin Series*, April 5, 1947, pp. 1–2; *Milwaukee Journal*, May 7, 21, 22, June 4, 20, 1947; *Milwaukee Sentinel*, May 21, June 4, July 24, 25, 1947.

27. *State ex rel. Drage* v. *Jones* 174 N.E. 783 (1930); *Mell* v. *State ex rel. Fritz*, 199 N.E. 72 (1935); *State ex rel. Hanrahan* v. *Zupnik*, 117 N.E.2d 689 (1954); *Thatcher* v. *Hogan*, 121 N.E.2d 130 (1954); *Roach* v. *State Board of Retirement*, 116 N.E.2d 850 (1954); *McCarthy* v. *State Board of Retirement*, 116 N.E.2d 852 (1954); *Smolinski* v. *Boston Retirement Board*, 190 N.E.2d 877 (1963); *Opinion of the Justices*, 303 N.E.2d 320 (1973); Annot., Vested Right of Pensioner to Pension, 52 A.L.R. 437, pp. 461, 471–74.

28. Annot., Vested Right of Pensioner to Pension, 52 A.L.R.2d 437. See also the A.L.R.'s *Later Case Service*, the supplement to the above.

29. *Pennie* v. *Reis*, quotes on pp. 470–71; Rubin G. Cohn, "Public Employee Retirement Plans—The Nature of the Employees' Rights," *University of Illinois Law Forum*, 1968, pp. 34–36.

30. *State ex rel. Risch* v. *Board of Trustees of Policemen's Pension Fund; Hughes*

v. *Traeger*, 106 N.E. 431 (1914), quote on p. 433; *Dodge* v. *Board of Education of Chicago*, 320 U.S. 74 (1937); *Ball* v. *Board of Trustees of Teachers' Retirement Fund*, 58 A. 111 (1904); Cohn, "Public Employee Retirement Plans," pp. 36–37.

31. *O'Dea* v. *Cook; Bender* v. *Anglin*, quote on p. 760; *Bakenhus* v. *City of Seattle; Yeazell* v. *Copins*, quote on p. 543; John F. Dillon, *Commentaries on the Law of Municipal Corporations* (Boston, 1911), vol. 1, p. 753; John D. Hicks, "The Constitutions of the Northwest States," *University [of Nebraska] Studies*, January–April 1923, pp. 1–162.

32. *Hickey* v. *Pension Board of City of Pittsburgh*, quotes on p. 235; Cohn, "Public Employee Retirement Plans," pp. 37–38.

33. *O'Dea* v. *Cook*, quotes on p. 367; *Retirement Board of Allegheny County* v. *McGovern*, quotes on pp. 404–5; *State ex rel. Gorczyca* v. *City of Minneapolis*, 219 N.W. 924 (1928); *Adamson* v. *City of Little Rock; Bakenhus* v. *City of Seattle*.

34. Cohn, "Public Employee Retirement Plans," p. 35.

35. *Retirement Board of Allegheny County* v. *McGovern*, quotes on p. 408; *Trotzier* v. *McElroy; Bender* v. *Anglin*, quotes on p. 760.

36. *People ex rel. Albright* v. *Board of Trustees of Firemen's Pension Fund of and for City and County of Denver*, 82 P.2d 765 (1938); *Board of Trustees of Firemen's Pension Fund for City & County of Denver* v. *People ex rel. Behrman; Police Pension and Relief Board of City and County of Denver* v. *McPhail*, quotes on pp. 698–99.

37. David H. Rosenbloom, "The Constitution and the Civil Service: Some Recent Developments, Judicial and Political," *Kansas Law Review*, 1970, pp. 839–69; William W. Van Alstyne, "The Demise of the Right-Privilege Distinction in Constitutional Law," *Harvard Law Review*, May 1968, pp. 1439–64; "The Policeman: Must He Be a Second-Class Citizen with Regard to His First Amendment Rights?" *New York University Law Review*, May 1971, pp. 536–59.

38. *Kern* v. *City of Long Beach*, 179 P.2d 799, quote on p. 803; *Bender* v. *Anglin*, quote on p. 760; *Hickey* v. *Pension Board of City of Pittsburgh*, quote on p. 237.

39. *Hickey* v. *Pension Board of City of Pittsburgh*, quote on p. 235; *James* v. *Police Pension Commission*, 87 D.&C. 454 (1953); *Yeazell* v. *Copins; Retirement Board of Allegheny County* v. *McGovern*, quote on pp. 404–5; *Dryden* v. *Board of Pension Commissioners of the City of Los Angeles; Bakenhus* v. *City of Seattle*, quote on p. 539; *Bender* v. *Anglin; Baker* v. *Retirement Board of Allegheny County*, 97 A.2d 231 (1953); *Police Pension and Relief Board of City and County of Denver* v. *Bills*.

40. *Hickey* v. *Pension Board of City of Pittsburgh*, quotes on pp. 237–38; *Yeazell* v. *Copins*.

41. *Kern* v. *City of Long Beach*, quote on p. 803; *Allen* v. *City of Long Beach*, quote on p. 767; *Bakenhus* v. *City of Seattle; Police Pension and Relief Board of City and County of Denver* v. *Bills*.

42. *Bakenhus* v. *City of Seattle*, quote on p. 540; A.L.R., *Later Case Service*, pp. 348–49, 353, 355; *Lickert* v. *City of Omaha*, quote on p. 649; *Talbott* v. *Independent School Dist. of Des Moines*, 299 N.W. 556 (1941), quote on p. 563; *City*

of Dallas v. *Trammell; Creps* v. *Board of Firemen's Relief and Pension Fund Trustees of Amarillo; Spina* v. *Consolidated Police and Firemen's Pension Fund Commission*, quote on p. 175; Cohn, "Public Employee Retirement Plans," pp. 48–51.

43. *Bakenhus* v. *City of Seattle*, quotes on pp. 541–42, 544. See also *Yeazell* v. *Copins*, pp. 547–51.

44. *Muzquiz* v. *City of San Antonio*, 378 F. Supp. 949 (1974), quotes on pp. 955, 958.

45. Robert Tilove, *Public Employee Pension Funds* (New York: Columbia University Press, 1976), pp. 304–5; "Public Employee Pension Funds in Times of Fiscal Distress," *Harvard Law Review*, March 1977, pp. 998–99; *Opinion of the Justices; Yeazell* v. *Copins;* Levy Anderson, "Vested Rights in Public Retirement Benefits in Pennsylvania," *Temple Law Quarterly*, 1961, pp. 255–77.

46. *Retirement Board of Allegheny County* v. *McGovern*, quote on p. 402; *Hickey* v. *Pension Board of City of Pittsburgh*, quote on p. 238.

47. *Detroit Free Press*, October 26, 1968.

48. Interviews with Lawrence Martin, Minnesota Legislative Commission on Pensions and Retirement, July 28, 1980, and Stanley G. Peskar, League of Minnesota Cities, July 29, 1980.

49. "Possible Economies in New York City's Pension Costs," report of the Subcommittee on Pensions of the Mayor's Committee to Study City Finances (1943), p. 15, Municipal Reference Library, New York.

6. *Proposals for Change*

1. U.S. Congress, Senate Committee on Governmental Affairs, Subcommittee on Governmental Efficiency and the District of Columbia, *Hearings [on the] District of Columbia Pension Reform Act* (Washington, D.C., 1978), pp. 85–91; U.S. Congress, House Committee on Education and Labor, *Pension Task Force Report on Public Employee Retirement Systems* (Washington, D.C., 1978), pp. 129–35, 138–39; Robert Tilove, *Public Employee Pension Funds* (New York: Columbia University Press, 1976), pp. 203–5.

2. Tilove, *Public Employee Pension Funds*, pp. 131–41, 211–16, 348–52.

3. C. Erwin Piper to Personnel and Labor Relations Committee, report dated January 5, 1979, p. 12, Los Angeles City Council Files, no. 77-2826, Municipal Records Center, Los Angeles; Los Angeles Board of Pension Commissioners, *Annual Report: 1979*, p. 38; U.S. Bureau of the Census, *Finances of Employee-Retirement Systems of State and Local Governments in 1977–78* (Washington, D.C., 1979), p. 24; Martin E. Segal Company, "Colorado Police and Firemen's Pension Plans Actuarial Valuations as of January 1, 1978" (1979), pp. 60, 65, 67, 72, 75, 80; Conrad M. Siegel, Inc., "Act 293 Report. Municipal Pension Funds. Actuarial Study Analysis" (1977), schedules 4PCI, 4FCI; Detroit Policemen and Firemen Retirement System, *Thirty-Seventh Annual Report: 1978*, pp. 3, 4, 7.

4. Moody's Investors Service, *Moody's Industrial Manual* (New York, 1980), vol. 1, p. a49.

5. Los Angeles Board of Pension Commissioners, *Annual Report: 1960*, pp. 6,

30, 35, *1979*, pp. 7, 38; Los Angeles Department of Pensions, "Fire and Police Pension System" (1975), p. 8, Department of Pensions Files, Los Angeles; telephone conversation with William S. Hutchison, Los Angeles Fire and Police Pension System, May 4, 1981.

6. Legislative Retirement Study Commission, *Report to the 1969 Legislative Session of the State of Minnesota* (1969), pp. 30–34; U.S. Congress, Senate Committee on the District of Columbia, *Hearings [on] Fiscal Pressures on the District of Columbia: Part 3. Pension Systems* (Washington, D.C., 1976), pp. 439–42; Harry Charles Katz, "The Impact of Public Employee Unions on City Budgeting and Employee Remuneration—A Case Study of San Francisco" (Ph.D. diss., University of California, Berkeley, 1977), pp. 131–32.

7. Office of the State Actuary, "Law Enforcement Officers and Fire Fighters Retirement System of the State of Washington Actuarial Valuation as of December 31, 1978" (1980), pp. 44–51; "Summary of the Major Provisions of the District of Columbia Retirement Reform Act (P.L. 96-122)," pp. 1–2, Office of the City Administrator's Files, Washington, D.C.; Dick Brown, "Policemen's and Firemen's Pensions: 1978 and 1979 Legislative Reforms" (1979), pp. 7–10; telephone conversation with William S. Hutchison, Los Angeles Fire and Police Pension System, November 17, 1980. The changes applied only to new appointees in all the cities referred to except Washington, D.C., where some of them applied to current employees as well.

8. New York City Commission on Pensions, *Report on the Pension Funds of the City of New York* (New York, 1918), pt. 3, p. 2; Municipal League of Los Angeles, *Bulletin*, August 28, 1925, p. 6; Tom Ingersoll to Los Angeles Board of Pension Commissioners, November 30, 1931, Department of Pension Files; *Detroit News*, February 6, 1936; New Orleans Bureau of Governmental Research, *City Problems*, May 22, 1944, p. 430.

9. *New Orleans States-Item*, May 3, 4, June 10, 1967; "Summary of the Major Provisions of the District of Columbia Retirement Reform Act," pp. 1–2; Hutchison telephone conversation, November 17, 1980; Senate Subcommittee on Governmental Efficiency and the District of Columbia, *Hearings on the District of Columbia Pension Reform Act*, pp. 42–43.

10. New Orleans Bureau of Governmental Research, *City Problems*, May 22, 1944, pp. 428–29; Los Angeles Board of Pension Commissioners, *Annual Report: 1979*, p. 14.

11. U.S. Congress, House Committee on Ways and Means, Subcommittee on Social Security, *Hearings [on] Coverage and Termination of Coverage of Government and Nonprofit Organization Employees Under the Social Security System* (Washington, D.C., 1976), pp. 196, 205; Senate Committee on the District of Columbia, *Hearings on Fiscal Pressures on the District of Columbia*, pp. 359, 381, 396, 418–20.

12. Senate Committee on the District of Columbia, *Hearings on Fiscal Pressures on the District of Columbia*, pp. 30–31, 396, 443–44; New York City Commission on Pensions, *Report on the Pension Funds of the City of New York* (New York, 1916), pt. 1, p. 124; annual reports of the New York City Police Pension Fund, 1925–32; District of Columbia Personnel Office, Compensation and Re-

search Division, "Total Police and Fire Retirements by Year for the Past 12 Years with Number and Percent for Optional (Longevity) and Disability," table dated December 1977, District of Columbia Personnel Office Files, Washington, D.C.

13. *New York Times*, May 21, 1980; *Washington Post*, January 12, February 26, 1978; Bureau of the Census, *Finances of Employee-Retirement Systems of State and Local Governments in 1977–78*, p. 36; *San Francisco Examiner*, November 29, 1979; San Francisco City and County Employees' Retirement System, *Annual Report: 1978–1979*, p. 35; *Seattle Post-Intelligencer*, January 24, 1977. New York City figures are for the post-1940 systems, Article II and Article IB.

14. Bureau of the Census, *Finances of Employee-Retirement Systems of State and Local Governments in 1977–78*, pp. 30–38; U.S. Bureau of the Census, *Census of Governments, 1977, Volume 6, Topical Studies: Number 1, Employee-Retirement Systems of State and Local Governments* (Washington, D.C., 1978), pp. 30–55. Figures for firemen and policemen are for 1978 except in the cases of Detroit and, for policemen, Denver, where they are for 1977. Figures for nonuniformed employees are for 1978 except in the cases of New York City, where they are for 1977, and Denver, where they are for 1976. New York City figures are for the post-1940 systems, Article II and Article IB.

15. Givens and Shafer, "Report of Examination of the Fire and Police Disability and Retirement Fund City of Portland, Oregon as of June 30, 1957" (1957), exhibit L; U.S. Bureau of the Census, *1957 Census of Governments: Volume IV, Topical Studies: Number 1, Employee-Retirement Systems of State and Local Governments* (Washington, D.C., 1959), pp. 30–49; U.S. Bureau of the Census, *1972 Census of Governments: Volume 6, Topical Studies: Number 1, Employee Retirement Systems of State and Local Governments* (Washington, D.C., 1973), pp. 30–53; Bureau of the Census, *Finances of Employee Retirement Systems of State and Local Governments in 1977–78*, pp. 30–38; Richard F. Camus and Associates, "Actuarial Study of the New Orleans Fire and Police Pension and Relief Funds as of June 30, 1977" (1978), pp. 42F, 43F, 24P, 25P.

16. Bureau of the Census, *Employee Retirement Systems of State and Local Governments: 1972*, pp. 30–53; Bureau of the Census, *Finances of Employee-Retirement Systems of State and Local Governments in 1977–78*, pp. 30–38; District of Columbia Personnel Office, "Total D.C. Police and Fire Retirements by Year."

17. M. Lewis Thompson, "Disability Pensions: No Problem? Big Problem? or Rip-off?" paper delivered at the Thirty-Seventh Annual Meeting of the National Conference on Public Employee-Retirement Systems (1978), p. 2; *Baltimore Sun*, April 22, 1974, May 11, 1975; Bureau of the Census, *Employee Retirement Systems of State and Local Governments: 1972*, pp. 30–53; Bureau of the Census, *Finances of Employee-Retirement Systems of State and Local Governments in 1977–78*, pp. 30–38.

18. *San Francisco Examiner*, November 26, 1979; *Washington Post*, January 12, 1978; U.S. Congress, House Subcommittee on Fiscal Affairs and Committee on the District of Columbia, *Hearings and Markups [on] Financing Retirement Funds for Police, Firemen, Teachers, and Judges* (Washington, D.C., 1976), p. 280; San Francisco Chamber of Commerce, "Study of the City Employee Retirement System City and County of San Francisco" (1977), pp. 15–16; Senate Subcommittee

on Governmental Efficiency and the District of Columbia, *Hearings on the District of Columbia Pension Reform Act*, pp. 44–53, 64–71; Thompson, "Disability Pensions," pp. 13–14; *Frye v. United States*, 72 F. Supp. 405 (1947); *Philadelphia Inquirer*, September 17, 1972. San Francisco figures are for the new plan. Under the old plan, a widow is entitled to 75 percent of an officer's allowance if he was on a service pension and 100 percent if he was on a disability pension.

19. *U.S. News & World Report*, October 17, 1971, pp. 104–6; Bureau of the Census, *Employee Retirement Systems of State and Local Governments: 1972*, pp. 30–53; Bureau of the Census, *Finances of Employee-Retirement Systems of State and Local Governments in 1977–78*, pp. 30–38; Piper to Personnel and Labor Relations Committee, pp. 21–22; *Philadelphia Inquirer*, September 17, 1972; *Baltimore Sun*, April 22, 1974.

20. House Committee on Education and Labor, *Pension Task Force Report*, pp. 206–7; Office of the State Actuary, "Report on Disability Retirements in the Law Enforcement Officers and Firefighters Retirement System" (1980), pp. 9, 50; *San Francisco Examiner*, November 28, 29, 1979.

21. Pennsylvania Crime Commission, *Report on Police Corruption and the Quality of Law Enforcement in Philadelphia* (1974), p. 543: Thompson, "Disability Pensions," p. 14; *Boston Globe*, May 16, 1976; *San Francisco Examiner*, November 27, 29, 30, 1979; Office of the State Comptroller, Office of the Special Deputy Comptroller for New York City, "New York City Fire Department Pension Fund: A Fiscal Analysis of Accidental Disability Retirement Benefits" (1979), pp. 22–26; *New York Times*, April 21, 1976; *Baltimore Sun*, December 24, 1975; *Minneapolis Tribune*, September 22, 1978; San Francisco Chamber of Commerce, "Study of the City Employee Retirement System," pp. 16–17; U.S. Congress, House Subcommittee on Revenue and Financial Affairs and Committee on the District of Columbia, *Hearings [on] Police and Firemen's Pay and Retirement* (Washington, D.C., 1974), p. 17.

22. Office of the State Comptroller, "New York City Fire Department Pension Fund," pp. vii–viii; San Francisco Chamber of Commerce, "Study of the City Employee Retirement System," pp. 17–18; *Baltimore Sun*, May 11, 1975; Thompson, "Disability Pensions," pp. 17–18; Office of the State Actuary, "Report on Disability Retirements," pp. 59–67.

23. Office of the State Actuary, "Law Enforcement Officers and Firefighters Retirement System Actuarial Valuation as of December 31, 1978," pp. 44–51; *New York Times*, May 21, 1980; "Summary of the Major Provisions of the District of Columbia Retirement Reform Act," pp. 1–2; *Washington Post*, January 18, 1978. The changes in the LEOFF system apply only to officers appointed on or after October 1, 1977. The retirement allowances are also reduced if an officer is less than fifty-eight years old.

24. Letter to the author from Jonathan Schwartz, New York City Employees' Retirement System, December 17, 1980; Los Angeles Board of Pension Commissioners, *Annual Report: 1979*, pp. 14, 17.

25. Los Angeles Board of Pension Commissioners, *Annual Report: 1979*, pp. 14, 17; Senate Committee on the District of Columbia, *Hearings on Fiscal Pressures on the District of Columbia*, p. 290; Schwartz letter, December 17, 1980;

Oakland Police and Fire Retirement System, *Twenty[-]Fourth Annual Report: 1975*, pp. 29, 30. The figures for New York City are for service-connected disability exclusively.

26. J. Douglas Brown, *Essays on Social Security* (Princeton: Industrial Relations Section, Princeton University, 1977), ch. 5; Tilove, *Public Employee Pension Funds*, p. 117; Robert J. Myers, "Should State and Local Governments Desert the Social Security Ship?" *Tax Review*, November 1976, p. 40; Neal R. Pierce and Jerry Hagstrom, "Some New Remedies for Public Pensions' Financial Ills," *National Journal*, November 26, 1977, pp. 1847–48; President's Commission on Pension Policy, *An Interim Report* (1980), p. 37; Joseph G. Metz, "The Case for Direct Integration with Social Security: New York State's Coordinated Escalator Retirement Plan," *Governmental Finance*, February 1978, pp. 26–29.

27. Tilove, *Public Employee Pension Funds*, pp. 117–19; Brown, *Essays on Social Security*, pp. 64–66; House Subcommittee on Social Security, *Hearings on Coverage and Termination of Coverage*, pp. 5–6, 10–11.

28. Joseph Krislov, "Extension of OASDI to State and Local Government Employees," *Public Personnel Review*, October 1957, pp. 213–14; Pierce and Hagstrom, "Some New Remedies for Public Pensions' Financial Ills," p. 1848; Advisory Commission on Intergovernmental Relations, *State and Local Pension Systems: Federal Regulatory Issues* (Washington, D.C., 1980), pp. 12–15; President's Commission on Pension Policy, *An Interim Report*, pp. 34–37.

29. Tilove, *Public Employee Pension Funds*, pp. 96–97; House Subcommittee on Social Security, *Hearings on Coverage and Termination of Coverage*, pp. 5–7; House Committee on Education and Labor, *Pension Task Force Report*, pp. 57–58.

30. U.S. Congress, Senate Committee on Finance, *Hearings [on] Social Security Revision* (Washington, D.C., 1950), pt. 2, pp. 881–86, 937–40, 964–68, 1125, pt. 3, pp. 1168–80; U.S. Congress, Senate Committee on Finance, *Hearings [on] Social Security Amendments of 1954* (Washington, D.C., 1954), pp. 234–38, 268–78; U.S. Congress, Senate Committee on Finance, *Hearings [on] Social Security Amendments of 1967* (Washington, D.C., 1967), pt. 2, pp. 1057–60; House Subcommittee on Social Security, *Hearings on Coverage and Termination of Coverage*, pp. 194–209.

31. House Subcommittee on Social Security, *Hearings on Coverage and Termination of Coverage*, pp. 114–18, 266–69; Senate Subcommittee on Governmental Efficiency and the District of Columbia, *Hearings on the District of Columbia Pension Reform Act*, pp. 85–91; Tilove, *Public Employee Pension Funds*, pp. 233–34.

32. Senate Finance Committee, *Hearings on Social Security Amendments of 1954*, pp. 235–36.

33. Metz, "The Case for Direct Integration with Social Security," pp. 26–29; "Recommendation for a New Pension Plan for Public Employees: The 1976 Coordinated Escalator Retirement Plan," report of the New York State Permanent Commission on Public Employee Pension and Retirement Systems (1976), pp. 1–59; House Subcommittee on Social Security, *Hearings on Coverage and Termination of Coverage*, pp. 91–92; Jonathan Schwartz to Kenneth Altman, memorandum dated December 31, 1976, a copy of which was made available to me by

Mr. Schwartz, chief actuary of the New York City Employees' Retirement System; Emily Jean Tourin, "The Politics of Pension Reform: New York State 1971–76" (Master's thesis, MIT, 1979).

34. Civic Federation of Chicago, *Bulletin No. 82*, April 1927; Katz, "Impact of Public Employee Unions," p. 131; Office of the State Actuary, "Law Enforcement Officers and Firefighters Retirement System of the State of Washington Actuarial Valuation as of December 31, 1978," pp. 48, 51; Hutchison telephone conversation, November 17, 1980; interview with Lawrence Martin, Minnesota Legislative Commission on Pensions and Retirement, July 28, 1980. The changes in the cost-of-living escalators applied only to new members.

35. House Committee on Education and Labor, *Pension Task Force Report*, pp. 258–59.

36. President's Commission on Pension Policy, *Coming of Age: Toward a National Retirement Income Policy* (1981), p. 32: *New York Times*, May 18, 1980; William C. Greenough and Francis P. King, *Pension Plans and Public Policy* (New York: Columbia University Press, 1976), pp. 74–75, 201–02; Robert J. Myers, *Indexation of Pension and Other Benefits* (Homewood, Ill.: Richard D. Irwin, 1978), chs. 3 and 4.

37. U.S. Public Health Service, *Vital Statistics of the United States, 1973. Volume II-Section 5. Life Tables* (Rockville, Md., 1975), p. 14; President's Commission on Pension Policy, *Coming of Age*, p. 32.

38. Greenough and King, *Pension Plans and Public Policy*, pp. 74–75.

39. Public Health Service, *Life Tables*, p. 14.

40. Senate Committee on the District of Columbia, *Hearings on Fiscal Pressures on the District of Columbia*, pp. 232–33.

41. *City of Passaic* v. *Consolidated Police and Firemen's Pension Fund Commission*, 113 A.2d. 22; Ohio Legislative Service Commission, *Police and Firemen's Pension Funds* (Columbus, 1965), pp. 45–47; New Orleans Bureau of Governmental Research, "The Retirement Systems for City of New Orleans Employees" (1974), pp. 2–3; Legislative Council Staff to Committee on Fire and Police Pensions, "Background on Fire and Police Pensions," memorandum dated July 18, 1977, pp. 1–10, 17–18, Colorado Municipal League Files, Wheat Ridge, Colorado; Legislative Commission on Pensions and Retirement, *Report to the 1979–1980 Minnesota State Legislature* (1980), p. 6.

42. Citizens' Bureau of Milwaukee, "How Other Cities Established a Single Pension System for Firemen, Policemen, and General City Employees" (1945), pp. 1–2.

43. Emma Schweppe, *The Firemen's and Patrolmen's Unions in the City of New York* (New York: Columbia University Press, 1948), p. 187.

44. John Mandeville to William N. Kelly, "Advantages and Disadvantages of Phasing Out Local Police and Paid Fire Funds Into PERA P&F," memorandum dated July 9, 1976, pp. 1–5, Minnesota Legislative Commission on Pensions and Retirement Files, St. Paul.

45. Interview with Jerry C. Kempf, Denver Manager of Revenue, January 15, 1981.

46. Interviews with Mayor Lloyd K. Poissenot, Sr., New Orleans Police De-

partment, and Bernard V. Nicolay, New Orleans Fire Fighters' Pension and Relief Fund, February 13, 1981.

47. *City of Passaic* v. *Consolidated Police and Firemen's Pension Fund Commission;* Ohio Legislative Service Commission, *Funding of the Police and Firemen's Disability and Pension Fund* (Columbus, 1969), p. 17; Poissenot interview, February 13, 1981; Nicolay interview, February 13, 1981; Brown, "Policemen's and Firemen's Pensions," pp. 4–10; Legislative Commission on Pensions and Retirement, *Report to the 1979–1980 Minnesota State Legislature,* p. 6.

48. Los Angeles Board of Pension Commissioners, *Annual Report: 1980,* p. 39.

49. Office of the State Actuary, "Law Enforcement Officers and Firefighters Retirement System Actuarial Valuation as of December 31, 1978," pp. 44–51.

50. Schweppe, *Firemen's and Patrolmen's Unions in the City of New York,* chs. 2–7; *Seattle Times,* January 9, 1977.

51. Office of the State Actuary, "Law Enforcement Officers and Firefighters Retirement System Actuarial Valuation as of December 31, 1978," p. 44; Legislative Commission on Pensions and Retirement, *Report to the 1979–1980 Minnesota State Legislature,* pp. 7–8; John Blackmore, "Pensions: Something Has Got to Give," *Police Magazine,* May 1978, p. 15.

52. Thomas P. Bleakney, "Washington Law Enforcement Officers' and Fire Fighters' Retirement System First Actuarial Valuation" (1970), pp. 2, 6; Thomas P. Bleakney, "Washington Law Enforcement Officers and Fire Fighters Retirement System Actuarial Valuation as of December 31, 1975" (1976), pp. 2, 3, 9, 46–47; *Seattle Post-Intelligencer,* January 26, 1977; *Seattle Times,* April 14, 1974; Office of the State Actuary, "Law Enforcement Officers' and Fire Fighters' Retirement System Actuarial Valuation as of December 31, 1978," pp. 2, 44–51.

53. U.S. Congress, Senate Committee on Government Operations, Subcommittee on Executive Reorganization, *Hearings [on the] Federal Role in Urban Affairs* (Washington, D.C., 1966), pt. 3, p. 774.

Conclusion

1. Martin E. Segal Company, "Boston Police Department Special Report on Retirement" (1974), p. 34.

2. Town Hall of California, *The Pension Balloon* (Los Angeles, 1979), pp. 37–46.

3. *Los Angeles Times,* June 3, 10, 1982.

4. *Ibid.,* May 27, 1982.

5. U.S. Congress, House Committee on Education and Labor, *Pension Task Force Report on Public Employee Retirement Systems* (Washington, D.C., 1978), pp. 232–37.

6. U.S. Congress, House Committee on Ways and Means, Subcommittee on Social Security, *Hearings [on] Coverage and Termination of Coverage of Government and Non-Profit Organization Employees Under the Social Security System* (Washington, D.C., 1976), p. 196.

7. *Ibid.*

Index

Bei Fragen zur Produktsicherheit wenden Sie sich bitte an:
If you have any questions regarding product safety,
please contact:

Walter de Gruyter GmbH
Genthiner Straße 13
10785 Berlin
productsafety@degruyterbrill.com